C0-ALP-026

The Determination and Use of Stability Constants

Arthur E. Martell
Ramunas J. Motekaitis
Texas A & M University

Arthur E. Martell
Ramunas J. Motekaitis
Chemistry Department
Texas A & M University
College Station, TX 77843-3255

Library of Congress Cataloging-in-Publication Data

Martell, Arthur Earl, 1916-
The determination and use of stability constants.

Bibligraphy: p.
Includes index.
1. Chemical equilibrium. I. Motekaitis, Ramunas J.
II. Title.
QD503.M374 1988 541.3'92 88-33750
ISBN 0-89573-715-9

Printed in the United States of America.

ISBN 0-89573-741-8 VCH Publishers
ISBN 3-527-26961-4 VCH Verlagsgesellschaft

Distributed in North America by:

VCH Publishers, Inc.
220 East 23rd Street, Suite 909
New York, New York 10010

Distributed Worldwide by:

VCH Verlagsgesellschaft mbH
P.O. Box 1260/1280
D-6940 Weinheim
Federal Republic of Germany

TABLE OF CONTENTS

PREFACE

The use of computers has revolutionized the calculation of stability constants from equilibrium data so that now very complex systems may be handled with relative ease. This new expertise, however, has produced problems attendant with the proliferation of sometimes meaningless parameters resulting from a seeming compulsion on the part of some investigators to go to almost any length to fit the experimental data. Because of such possibilities considerable care and insight into the principles of coordination chemistry are needed to avoid the postulation of unjustified species and the publication of meaningless equilibrium constants. In addition to explaining the computational methods for determining stability constants considerable attention will be devoted in this book to safeguards and methods for avoiding mis-interpretation and over-interpretation of equilibrium data. The use of computers has also made possible the development of various methods of displaying experimental results, such as the production of species distribution diagrams. Such graphical illustrations may be used to test the self-consistency of the calculated results, the validity of the data used for computer processing, as well as the assumptions and methodology employed for defining the chemical system under investigation. These new developments will be described and will be illustrated with examples of several studies of complex systems from the authors' laboratory.

The large amount of stability data that have been produced in various parts of the world has resulted in wide diversity in both the nature and quality of the publications in this field. Because of the proliferation of poor or ambiguous data, the leading investigators of complex equilibria have set up quality standards for measuring and reporting stability data, and proposed that they be enforced by the editors of journals dealing with papers in this field. The proposed standards, which are similar to those employed in developing the series of books on Critical Stability Constants, by Martell and Smith, are described.

Several multicomponent systems are also described to illustrate the use of equilibrium constants for the elucidation of complex systems involving several multidentate ligands and a number of metal ions. Descriptions are provided for the development and use of a comprehensive data base applicable to the determination of individual molecular species in multicomponent systems of the type encountered in the environment, in biological fluids, and other complex natural systems.

Because the experimental data processing techniques now employed in this field are not usually taught in college and university curricula, and are generally unknown to U.S. chemistry majors at all degree levels, it is hoped that this small book will serve as a convenient manual for those beginning work with metal complexes in solution, and to establish uniform guidelines for those already engaged in the measurement and use of metal complex stability constants.

ACKNOWLEDGEMENT

The authors acknowledge with thanks the assistance of Dr. Janet Marshall in reviewing an early version of the manuscript, and of Mary Martell for preparation of the photo-ready copy.

The Determination and Use of Stability Constants

INTRODUCTION

1.1 Stability Constants - Early Work

Stability constants, or equilibrium constants for metal complex formation, have long been employed as an effective measure of the affinity of a ligand for a metal ion in solution, and have served as a quantitative indication of the success or failure of ligand design. The earliest quantitative determinations were concerned only with the determination of empirical formulas and overall formation constants, and were pioneered by workers such as von Euler[1] and Bodlander.[2] The stepwise hydrolysis constants of Cr(III) were reported by N. Bjerrum.[3] The measurement of stepwise stability constants for monodentate ligands effectively began with the classic dissertation of J. Bjerrum[4] on the formation of transition metal-ammonia complexes in aqueous solution. Stability work on chelate compounds began with the seminal paper by Calvin and Wilson[5] wherein the Bjerrum[4] simplification of using a large excess of ligand to prevent hydrolysis and precipitation was avoided. This work was the beginning of a trend involving the use of exact algebraic treatment of equilibrium constants and mass balance equations, thus eliminating the approximations of the Bjerrum method. Because of the power of these methods of determining formation constants of complexes, and the relatively inexpensive instrumental requirements, work in this field rapidly proliferated. The results are collected in a series of extensive non-critical compilations,[6-10] now sponsored by the Subcommission V6 on Equilibrium Data of IUPAC. A series of Critical Tables of Stability Constants has been published recently.[11] The Commission on Equilibrium Data has also sponsored a series of in depth critical reviews of stability constants of individual ligands or groups of related ligands, and some of these have appeared.[12]

The proliferation of publications in this field, by which investigators would report stability constants for their own sake, without any apparent objective of advancing the concepts and principles of coordination chemistry, or of providing information essential to other fields, has led to a general decline in the prestige of this research area, and work on stability constants gradually came to be regarded as routine. This image has been further aggravated by the appearance of a large number of poor papers from developing countries and elsewhere, whereby a minor change in ligand structure was taken as the sole justification for yet another publication, and in which controls and conditions were neglected, resulting in the reporting of poor data. This proliferation of routine publications was made even worse by a recent flood of reports of mixed ligand constants, in which stability constants of two or more ligands with a single metal ion were studied for no apparent reason except to publish more constants. With thousands of ligands available, this process could go on forever, without appreciable contribution to the field.

1.2 Recent Developments

Three important, far-reaching developments have occurred recently in solution coordination chemistry which provide a major role for stability constant determinations and related information in the further development of the field. These new advances are 1, the development of the chemistry of macrocyclic and macrobicyclic ("cryptand") complexes, with the accompanying new challenges to ligand design and synthesis, and potential achievements in the study of new complexes with novel properties, high stabilities and important applications; 2, the development of the new fields of bioinorganic chemistry and inorganic environmental chemistry, both of which require knowledge of the complexes formed in multicomponent systems containing many ligands and metal ions; and 3, the development of computational methods for processing equilibrium data to provide more accurate and more rapid determination of stability constants, and the extension of the method to multidentate ligands, and to systems of many metal ions and ligands, that are too complex to have been investigated previously by classical methods. Now stability constants may be used, with the aid of appropriate computer programs, for the elucidation of the molecular and ionic species present in very complex biological and environmental systems.

1.3 Historical Evolution of Computational Methods

Although methods of determining stability constants have been described in several books[13,14,15] the monograph by Rossotti and Rossotti[16] still stands today as the authority on pre-computer methods of data reduction involving many types of equilibrium data. It ties together definitions of equilibrium constants with derivations of graphical methods available at the time. Because of the complexity of the systems containing even relatively few species, approximation techniques were the rule rather than the exception.

With the advent of the computer, the situation changed dramatically. With this new tool, the equilibrium coordination chemist became able to include in his treatment of data all of the equilibrium constants simultaneously. However, as Gaizer[17] pointed out in his 1979 review, although computer procedures had been described in over 100 publications, papers were still appearing in which authors either avoided use of the computer altogether, or used it only to diminish the labor of the same tedious hand calculations which they had been carrying out for many years. This reluctance by cautious workers, although slowing down progress, was probably better than using a computer equilibrium program as a black box without regard to limitations of the computer program or limitations imposed by the data. From a mathematical viewpoint, the review by Gans[18] is an excellent survey of the field and the pitfalls in the application of computers to experimental data.

A number of computer programs are currently available[17,18,19] which employ varying strategies[20,21] and new programs are being published with regularity. The classical programs have been compared,[22,23] but no single program seems to have outstanding advantages over the others. Each of the published programs differs at least in some detail (see Table 1 in Ref. 21 for a typical comparison). The computer program selected for discussion herein is unique in that it is convenient, is especially adapted to the processing of potentiometric pH data, and calculates the pH data directly for comparison with experimental measurements.

1.4 Purpose of This Book

In this book examples will be described for the application of stability constant determinations to the elucidation of the complexing properties of macrocyclic and macrobicyclic complexes, and the detection of secondary ("cascade") binding of additional donors to the complexes initially formed. Examples are also given for the use of stability constants in calculating distributions and concentrations of metal species (the term "speciation" is often employed) in complex systems containing many ligands and metal ions, such as physiological fluids (e.g., blood serum) and environmental solutions (e.g., sea water). The main thrust, however, will be to provide a thorough description of the experimental requirements for obtaining experimental data of high quality, and to explain in detail the modern computational methods for determining stability constants from such data. This type of information does not seem to be available currently in a logical, systematic form, although fragmentary information is scattered throughout the literature. Such a treatment is certainly not readily available to students, because much of it has been dropped from academic curricula, especially in the U.S.A. and the United Kingdom. Because of the rapid new growth of this field, a clear detailed description of modern experimental and computational methods is now considered imperative.

In order to increase the usefulness of this book three Fortran 77 computer programs are included on the accompanying MS DOS diskette. PKAS is suited for the computation of overlapping protonation constants, BEST is the workhorse for the computation of potentiometric stability constants, SPE determines species distributions from known constants. The plotting routine for the latter is written in GW Basic. All the species distribution diagrams were generated on a HP LaserJet$^+$ or LaserJet Series II using SPE.

References

1. von Euler, H. Ber., **1903**, 1854.
2. Bodlander, G.; Storbeck, O. Z. Anorg. Chem., **1902**, 31, 1.
3. Bjerrum, N. Ph.D. Dissertation, Copenhagen, 1908.
4. Bjerrum, J. "Metal Amine Formation in Aqueous Solution"; Thesis, Copenhagen, 1941; reprinted 1957, P. Haase and Son, Copenhagen.
5. Calvin, M.; Wilson, K. W. J. Am. Chem. Soc., **1945**, 67, 2003.
6. "Stability Constants. Part I. Organic Ligands"; "Part II. Inorganic Ligands", Schwarzenbach, G.; Sillen L. G. Eds.; Chemical Society: London, 1957, 1958.
7. "Stability Constants, Part II, Organic Ligands" Special Publication No.17, Eds. Sillen, L. G.; Martell, A. E.; Chemical Society: London, 1964.
8. "Stability Constants, Supplement No.1", Special Publication No.25, Eds. Sillen, L. G.; Martell, A. E.; Chemical Society: London, 1971.

9. "Stability Constants of Metal-ion Complexes. Part B: Organic Ligands", Ed. D. D. Perrin; Pergamon Press: Oxford, 1979.
10. "Stability Constants of Metal-ion Complexes. Part A. Inorganic Ligands", Ed. E. Hogfeldt; Pergamon Press: Oxford, 1982.
11. Smith, R. M.; Martell, A. E. "Critical Stability Constants" Vol.1, 2, 3, 4, 5; Plenum: New York, 1974, 1975, 1977, 1976, 1982.
12. a) Anderegg, G. "Critical Survey of Stability Constants of EDTA Complexes"; Pergamon Press: Oxford, 1977; b) McBryde, W. A. E. "A Critical Review of Equilibrium Data for Proton and Metal Compounds"; Pergamon Press: Oxford, 1978; c) Stary, J.; Zolotov, Yu. A.; Petrukhin, O. M. "Critical Evaluation of Equilibrium Constants involving 8-Hydroxyquinoline and its Metal Chelates"; Pergamon Press: Oxford, 1979; d) Bond, A. M.; Hefter, G. T. "Critical Survey of Stability Constants and Related Thermodynamic Data of Fluoride Complexes in Aqueous Solution"; Pergamon Press: Oxford, 1980; e) Stary, J.; Liljenzin, J. D. Pure and Appld. Chem., **1982**, 54, 2557. f) Anderegg, G. Pure and Appld. Chem., **1982**, 54, 2693. g) Tuck, D. G. Pure and Appld. Chem., **1983**, 55 1477. h) Pettit, L. D. Pure and Appld. Chem., **1984**, 56, 247. i) Paoletti, P. Pure and Appld. Chem., **1984**, 56, 491. j) Beck, M. R. Pure and Appld. Chem., **1987**, 59, 1703.
13. Martell, A. E.; Calvin, M. "Chemistry of the Metal Chelate Compounds"; Prentice-Hall: New York, 1952.
14. Chaberek, S.; Martell, A. E. "Organic Sequestering Agents"; John Wiley: New York, 1959.
15. "Chelating Agents and Metal Chelates", Dwyer, F. P. and Mellor, D. P. Eds.; Academic Press: New York, 1964.
16. Rossotti, F. C.; Rossotti, H. "The Determination of Stability Constants"; McGraw-Hill: New York, 1961.
17. Gaizer, F. Coord. Chem. Revs., **1979**, 27 (3), 195.
18. Gans, F. Coord. Chem. Revs., **1976**, 19 (2), 99.
19. Rossotti, F. J. C.; Rossotti, H. S.; Whewell, R. J. J. Inorg. Nucl. Chem., **1971**, 33, 2051.
20. Izquierdo, A.; Beltran, J. L. Anal. Chim. Acta, **1986**, 181, 87.
21. Hofman, T.; Krzyzanowska, M. Talanta, **1986**, 33, 851.
22. Field, T. B.; McBryde, W. A. E. Can. J. Chem., **1978**, 56, 1202.
23. Legett, D. J. Talanta, **1977**, 24, 535.

EQUILIBRIUM CONSTANTS, PROTONATION CONSTANTS, FORMATION CONSTANTS

2.1 Concentration Constants and Activity Constants

An equilibrium constant is a quotient involving the concentrations or activities of reacting species in solution at equilibrium. It is often defined as the ratio of products of the activities of the reaction products, raised to the appropriate power, to the product of the activities of the reactants, raised to the appropriate power, illustrated by equation (2.1).

$$aA + bB \rightleftharpoons cC + dD \qquad K_{eq} = \frac{a_C^c \cdot a_D^d}{a_A^a \cdot a_B^b} \qquad (2.1)$$

The value of the equilibrium constant is directly related to the differences in the Gibbs free energies of products and reactants in their standard states (unit activity), and hence is a measure of the difference of reactivities of the reactants and products. With information on the enthalpy changes, the free energies may be broken down to heats and entropies of reaction, thus providing further insights.

The determination of the activities of complex ionic species at both infinite dilution and in real solutions is a complicated and time-consuming task, which is far beyond the interests and expertise of most coordination chemists. Because concentrations parallel activities of ionic solutes when the ionic strength is controlled by a non-reacting electrolyte present at a concentration far in excess (ca. 100 times) that of the reacting ionic species under investigation, it has become general practice from a practical point of view to measure equilibrium constants involving coordination compounds at constant ionic strength maintained by a supporting electrolyte. Indeed, historically this practice may be traced back to the turn of the century when Grossman[1] employed KNO_3 to keep ionic strength constant, although the formal introduction of the ionic strength concept came nearly two decades later in 1921.[2] The equilibrium constant determined at constant ionic strength would then be indicated by equation (2.2).

$$aA + bB \rightleftharpoons cC + dD \qquad K_c = \frac{[C]^c[D]^d}{[A]^a[B]^b} \tag{2.2}$$

where [ ] indicate molar concentrations

In addition to convenience of measurement, this practice has further advantages. For example the quantities involved may be substituted directly into the mass balance equations (see below) employed for stability constant and protonation constant calculations. Also the concentration quantities determined potentiometrically correspond to the quantities measured by other techniques (e.g., absorbance). The choice of the nature and concentration of the supporting electrolyte has become partially but not completely standardized, and this problem will be given some attention. For the physical chemists who have occasionally questioned the validity of the so-called "concentration constant", equation (2.2), it is pointed out that for a reference solution containing the supporting electrolyte, these constants are very close to the true "thermodynamic" constants because the activity coefficients change very little in going from solutions at finite concentration to solutions in which the solutes are at infinite dilution in the same supporting electrolyte.[3]

2.2 Conventions Employed for Expressing Equilibrium Constants

Protonation and complex formation equilibria may be expressed in several ways, and several "different" kinds are in common use in the literature, such as acid dissociation constants, stepwise dissociation constants, formation constants, instability constants, stepwise formation constants, overall formation constants, hydrolysis constants, displacement constants, exchange constants, isomerization constants, mixed ligand constants, mixed metal ion formation constants, etc.[3] The large number of equilibrium expressions employed indicates the importance of clearly defining the equilibrium quotients when publishing results in this field. The most commonly used expression in reporting results in the current literature are stepwise protonation constants for the ligand, and either overall metal complex protonation constants or stepwise formation constants.

Since this monograph deals primarily with aqueous equilibria involving coordination compounds, equilibrium constants will be described in terms of the combination of Lewis acid species, such as hydrogen ions and metal ions, with basic donor species (ligands). Also, for convenience in setting up a generalized algorithm for computational purposes the equilibria will be set up in terms of "overall" constants, designated by β, in place of the more popular and more readily visualized stepwise constants (K).

2.3 Equilibrium Constants and Stability Constants for EDTA

For clarity of presentation, a specific example, the equilibria between EDTA (with four ionizable hydrogens) and the Ca(II) ion, is presented below. EDTA may be represented by the symbolic formula H_4L. The four successive stepwise protonation constants of EDTA are represented by equations

(2.3)-(2.6). The relationship between the successive and the corresponding overall protonation constants (2.10-(2.13) is that the latter is a cumulative product of the former. Specifically, in this case, $\beta_{HL} = K_1^H$, $\beta_{H_2L} = K_1^H K_2^H$, $\beta_{H_3L} = K_1^H K_2^H K_3^H$ and $\beta_{H_4L} = K_1^H K_2^H K_3^H K_4^H$. In this example the calcium(II) formation constant, equation (2.7), is identical to the corresponding overall constant (2.14) while the overall protonated chelate constant (2.15) is a product of the stepwise constants (2.7) and (2.8): $\beta_{CaHL} = K_{CaL}^{Ca} K_{CaHL}^{H}$. Some of the species appearing in equations (2.3)-(2.9) are not present in the system over various parts of the pH range, but a complete set (2.3)-(2.8) or (2.10)-(2.15) is needed to describe the equilibria in the pH range over which hydrogen ion concentrations are measured. Equations (2.8) and (2.9) represent two ways of forming the protonated complex, and either equation may be used.

$$H^+ + L^{4-} \rightleftharpoons HL^{3-}$$

$$K_1^H = \frac{[HL^{3-}]}{[H^+][L^{4-}]} \quad (2.3) \qquad \beta_{HL} = \frac{[HL^{3-}]}{[H^+][L^{4-}]} \quad (2.10)$$

$$H^+ + HL^{3-} \rightleftharpoons H_2L^{2-}$$

$$K_2^H = \frac{[H_2L^{2-}]}{[H^+][HL^{3-}]} \quad (2.4) \qquad \beta_{H_2L} = \frac{[H_2L^{2-}]}{[H^+]^2[L^{4-}]} \quad (2.11)$$

$$H^+ + H_2L^{2-} \rightleftharpoons H_3L^-$$

$$K_3^H = \frac{[H_3L^-]}{[H^+][H_2L^{2-}]} \quad (2.5) \qquad \beta_{H_3L} = \frac{[H_3L^-]}{[H^+]^3[L^{4-}]} \quad (2.12)$$

$$H^+ + H_3L^- \rightleftharpoons H_4L$$

$$K_4^H = \frac{[H_4L]}{[H^+][H_3L^-]} \quad (2.6) \qquad \beta_{H_4L} = \frac{[H_4L]}{[H^+]^4[L^{4-}]} \quad (2.13)$$

$$Ca^{2+} + L^{4-} \rightleftharpoons CaL^{2-}$$

$$K^{Ca}_{CaL} = \frac{[CaL^{2-}]}{[Ca^{2+}][L^{4-}]} \quad (2.7) \qquad \beta_{CaL} = \frac{[CaL^{2-}]}{[Ca^{2+}][L^{4-}]} \quad (2.14)$$

$$H^{+} + CaL^{2-} \rightleftharpoons CaHL^{-}$$

$$K^{H}_{CaHL} = \frac{[CaHL^{-}]}{[H^{+}][CaL^{2-}]} \quad (2.8) \qquad \beta_{CaHL} = \frac{[CaHL^{-}]}{[H^{+}][Ca^{2+}][L^{4-}]} \quad (2.15)$$

$$K^{Ca}_{CaHL} = \frac{[CaHL^{-}]}{[Ca^{2+}][HL^{3-}]} \quad (2.9)$$

2.4 Species Distribution Curves

As a aid in visualizing the significance and implications of the equations (2.3)-(2.8), a so-called species distribution diagram for 1.00 x 10^{-3} M EDTA and 1.00 x 10^{-3} M Ca(II) was drawn up using Program SPE and is shown in Figure 2.1. As the p[H] is varied from 2 to 12, the solution changes from one containing uncomplexed Ca^{2+}, H_4L^{o}, H_3L^{-}, and a small but significant concentration of H_2L^{2-} to solutions which ultimately contain only CaL^{2-}. At pH ca. 4, the major species are about 0.50 x 10^{-3} M Ca^{2+}, CaL^{2-}, and H_2L^{2-} while in this region one also finds minor concentrations (ca. 3%) of $CaHL^{-}$ and HL^{3-}. Such diagrams are very useful not only for visualizing the nature of the equilibrium situation, but also are often helpful in raising questions about the logic of a published result. The fact that $CaHL^{-}$ appears with a maximum concentration of only 2.7% of the Ca(II) species present *a priori* casts some doubt on its existence or, at the very least, indicates that its concentration cannot be accurately known under the conditions employed, and that the β value calculated for this species has a relatively large uncertainty.

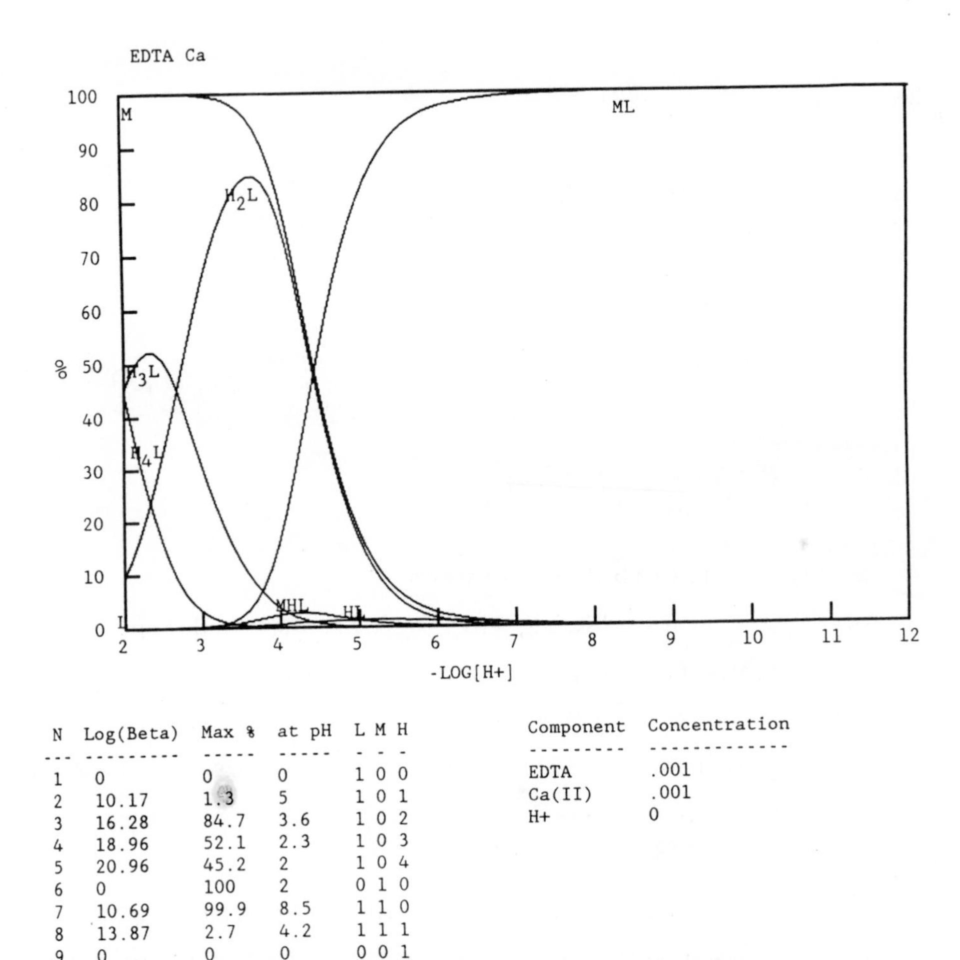

N	Log(Beta)	Max %	at pH	L M H
1	0	0	0	1 0 0
2	10.17	1.3	5	1 0 1
3	16.28	84.7	3.6	1 0 2
4	18.96	52.1	2.3	1 0 3
5	20.96	45.2	2	1 0 4
6	0	100	2	0 1 0
7	10.69	99.9	8.5	1 1 0
8	13.87	2.7	4.2	1 1 1
9	0	0	0	0 0 1
10	-13.78	0	0	0 0-1

Component	Concentration
EDTA	.001
Ca(II)	.001
H+	0

Figure 2.1. Species Distribution Diagram for 0.0010 M Ca(II) and 0.0010 M EDTA at 0.100 M Ionic Strength and 25.0°C.

References

1. Grossman, H. Z. Anorg. Chem., **1905**, 43, 356.
2. Lewis, G. N.; Randall, M. J. Am. Chem. Soc., **1921**, 43, 1140.
3. Rossotti, F. C.; Rossotti, H. "The Determination of Stability Constants"; McGraw-Hill: New York, 1961, p.17.

EXPERIMENTAL METHODS FOR MEASURING EQUILIBRIUM CONSTANTS

3.1 Methods Available

Any method which may be used to determine with reasonable accuracy the concentration of at least one of the species in equilibrium provides the information needed, together with the known analytical composition of the experimental solution, to calculate the concentrations of all of the remaining species present at equilibrium, and hence the equilibrium constant (eq.2.2). If a sufficient number of such equilibrium measurements are made over a sufficient range of conditions, accurate values of stability constants that apply within the range of reaction conditions employed can be obtained. A partial list of the methods that have been employed is presented in Table 3.1. Early treatises on the subject have described these standard methods in some detail.[1-4] Further discussion of the methods available may be found in a more recent review by Anderegg.[5]

Table 3.1. Methods for Determining Stability Constants

Standard Methods
Potentiometry
Spectrophotometry
Specific ion EMF measurements
Nuclear magnetic resonance spectroscopy
Polarography
Ion exchange
Colorimetry
Ionic conductivity
Distribution between two phases
Reaction Kinetics
Partial pressure measurement
Solubility measurement
Competition Methods for Strong Complexes
Metal-metal competition measured spectrophotometrically[6,7]
Ligand-ligand competition measured potentiometrically[7-10]
Ligand-ligand competition measured spectrophotometrically[8]
Special[9]

3.2 Potentiometric pH Measurements

Potentiometry is one of the most convenient and successful techniques employed for metal complex equilibrium measurements. While some workers measure metal ion concentration with specific ion electrodes, or with metal electrodes, it is usually sufficient to use the highly accurate glass electrode for measuring the hydrogen ion concentration in a procedure termed potentiometric titration, whereby, for example, standard base is added in increments to a well-characterized acid solution of the ligand in the absence of and in the presence of known (total) metal ion concentrations.

When complex formation is practically complete at the initial low pH, or proton-involving equilibria occur above pH 12, or when spectral changes indicate reactions that occur as the $p[H^+]$ ($-\log [H^+]$) is varied, electronic absorption spectrophotometry can be an important auxiliary technique for determining stability constants.

While there are many other methods (polarography, ion exchange, NMR, etc.)[5] their use is usually invoked under special circumstances which arise when potentiometry or spectrophotometry cannot be employed. For example ion exchange methods may be used for studying complex formation by trace concentrations of a radioactive metal ion. Potentiometry does not provide microscopic information involving identification of protonation and metal coordination sites on a ligand. For such information spectroscopic measurements (e.g., NMR) or spectrophotometric absorbance studies are needed.

3.3 Displacement Methods

The potentiometric measurement of hydrogen ion concentration may be employed when the degree of complex formation is sensitive to the hydrogen ion concentration, i.e., the degree of complex formation undergoes measurable change (increase) as the pH is changed (increased). When that does not occur, because the ligand is weakly basic and has little or no affinity for hydrogen ions, other methods must be employed. Also, in some cases the complex may be so stable that it is completely or nearly completely formed at the lowest measurable pH. Under such conditions substitution methods, such as those listed in Table 3.1, are sometimes available. Thus the metal ion may be displaced from the ligand by another metal[6] ion which has comparable affinity for the ligand (preferably within an order of magnitude of the stability constant, but certainly well within two orders of magnitude). The progress of the displacement can be followed if one of the metal ions, or its complex, has considerably different absorbance compared to the other.[7] Ligand-ligand displacement, followed spectrophotometrically, can be employed in an analogous manner.[8]

When the complex systems being investigated do not involve a color change or absorbance change, the displacement of one ligand by another may be followed potentiometrically if the sums of the protonation constants of the ligands are appreciably different.[9] Thus a very stable EDTA-transition metal complex may be converted to an even more stable complex of a polyamine (such as triaminotriethylamine), the extent of conversion being measured by the uptake of base required to neutralize the additional hydrogen ions displaced from the secondary ligand by the metal ion. Frequently a second metal ion which complexes the displaced ligand but not the secondary ligand, such as Ca^{2+}, may be added to provide an additional driving force for the displacement and to produce a simpler complex system after the displacement.[10] The equilibrium constant determined in this way corresponds to the displacement reaction 3.1 (charges omitted).

$$ML + H_nL' + M' \rightleftharpoons ML' + M'L + nH \qquad (3.1)$$

The calculation of the stability constant of ML requires that those of ML′ and M′L be known or readily measurable. This method was first employed by Schwarzenbach and coworkers[10] to determine the stability constants of transition metal ion-EDTA complexes.

A unique method developed by Motekaitis and Martell[11] takes advantage of the strong tendency of all Ga(III) complexes to be converted to $Ga(OH)_4^-$ at high pH. Because the stability constant of this complex is known in terms of Ga^{3+} and OH^- ions, it may be employed as the starting point of a potentiometric titration procedure whereby increments of standard acid solution are added to a solution of known p[H] (or p[OH]) initially consisting of $Ga(OH)_4^-$ and free ligand. An example of this procedure is described below. The method may prove to be useful for the investigation of other strongly amphoteric metal ions.

A problem that is frequently encountered with very stable complexes of multidentate ligands is the inability to neutralize, in the absence of metal ions, one or more of the donor groups of the ligand without raising the pH above 12, which is the practical upper limit for accurate pH measurements. While this does not affect the accuracy of the equilibrium determinations in the working pH range, 2-12, it renders impossible the expression of the stabilities in terms of the fully dissociated ligand, L^{n-}, and comparisons of such constants with the corresponding values of other metal complex systems cannot be made. It is sometimes possible to determine these high protonation constants by absorbance measurements, when there is appreciable and measurable difference in the absorbances of the fully deprotonated ligand and its protonated forms, as for ligands with very basic phenolate and catecholate donor groups. In such cases the hydrogen ion concentration is obtained from the analytical concentrations of hydroxide ion in the solution, since $[OH^-]$ far exceeds the concentrations of the ligand species, and accurate p[H] measurements are not possible. When such procedures are employed it must be realized that even though microscopic species are the source of the observed spectra, the absorbances in fact measure the macroscopic equilibria. It is only by analogy, usually involving approximations and assumptions, that microscopic equilibria may be obtained. These considerations are discussed in greater detail below under microscopic equilibria (Chapter 7).

3.4 pH and p[H]

Because of the widespread use of the term pH in potentiometric studies of ligand basicities and metal complex stabilities it is considered important to comment on the definition and measurement of this quantity. The quantity pH, defined as -log a_{H^+}, is not an explicitly measurable quantity, because it is not possible to determine the activity of a single positive or negative ion. The potential of the hydrogen electrode in the cell $H_2|H^+, Cl^-|AgCl|Ag$ is RT/F log $(a_{H^+} \cdot \gamma_{Cl^-})$ where γ_{Cl^-} is the activity coefficient of the chloride counter ion. In spite of this built-in restriction, the pH scale is nevertheless widely used as a measure of acidity or alkalinity of aqueous solutions. This should only be done, however, with the understanding that the values are approximate and that they cannot be precisely measured or explicitly defined. Under conditions of constant ionic strength maintained by an inert supporting electrolyte, activity coefficients are essentially constant, and the potential of the hydrogen electrode in a cell without liquid junction (e. g., H_2, $H^+|AgCl|Ag$; electrolyte = HCl, KCl) varies linearly with hydrogen ion concentration as well as with hydrogen ion activity. When the hydrogen electrode is replaced by a glass electrode (an ion selective electrode for H^+), and the electrolyte differs from that of the reference electrode, the parallel relationship remains, with additive constants for the liquid junction potentials, provided that the supporting electrolyte remains as the dominant ionic conductor. Below pH 2 and above pH 12, hydrogen ion and hydroxide ions begin to be responsible for appreciable fractions of the conductance, so that liquid junction potentials change as the pH is lowered and raised below and above the limits. A useful estimate of the effects of liquid junction potential on the expected accuracy of the pH measurement can be obtained from computations made by Bates.[12] For a saturated calomel electrode in contact with the experimental solution containing 0.100 M KCl through a liquid junction, E_j = 1.8 mV. On the other hand, if this electrode is in contact with 0.010 M HCl + 0.090 M KCl, E_j = 2.1, a mere 0.30 mV rise corresponding to a p[H] error of 0.005 units. Also, a solution of 0.010 M NaOH E_j = 2.3, but if $[OH^-]$ is only 10% of the background electrolyte, the change in the junction potential would be negligible. Accurate measurements of hydrogen ion concentration (or activity) with a glass electrode-reference electrode systems are restricted therefore to the range 2-12. For reasons which will be developed below, hydrogen ion concentrations will be employed exclusively in this book. In order to make this distinction clear, the generally used term pH is replaced by p[H], with the brackets indicating concentration.

References

1. Martell, A. E.; Calvin, M. "Chemistry of the Metal Chelate Compounds"; Prentice-Hall: New York, 1952.

2. Chaberek, S.; Martell, A. E. "Organic Sequestering Agents"; John Wiley: New York, 1959.
3. "Chelating Agents and Metal Chelates", Dwyer, F. P. and Mellor, D. P. Eds.; Academic Press: New York, 1964.
4. Rossotti, F. C.; Rossotti, H. "The Determination of Stability Constants"; McGraw-Hill: New York, 1961
5. Anderegg, G. Pure Appld. Chem., **1982**, 54, 2693.
6. Irving, H.; Mellor, D. H. J. Chem. Soc., **1955**, 3457.
7. Harris, W. R.; Martell, A. E. Inorg. Chem., **1976**, 15, 713.
8. Harris, W. R.; Motekaitis, R. J.; Martell, A. E. Inorg. Chem., **1975**, 14, 974.
9. Schwarzenbach, G.; Ackermann, H. Helv. Chim. Acta, **1949**, 32, 1543.
10. Schwarzenbach, G.; Freitag, E. Helv. Chim. Acta, **1951**, 34, 1492.
11. Motekaitis, R. J.; Martell, A. E. Inorg. Chem., **1980**, 19. 1646.
12. Bates, R. G. "Determination of pH"; John Wiley: New York, 1973; 2nd Ed, p.38.

EXPERIMENTAL PROCEDURE FOR POTENTIOMETRIC p[H] MEASUREMENT OF METAL COMPLEX EQUILIBRIA

For the purposes of discussion it is assumed that the ligand under consideration is highly basic and multidentate, and that the study involves determination of stability constants with a variety of metal ions at 25.0°C and with a supporting electrolyte of 0.100 M ionic strength.

4.1 Materials

It is important to assert that the ligand, as well as other reagents, must be of the highest purity possible. It would be best if the ligand were available in crystalline form, and that it had been recrystallized and characterized by NMR and by elemental analysis. If the ligand is available in sufficient quantity and stable in aqueous solution it may be prepared in the form of a stock solution (with a suggested molarity of 0.0100 to 0.0500). An alternate method, which may save material, is to weigh out the required amount for each run.

The metal ions to be studied may be made up as standard solutions of their salts (preferably as nitrates or perchlorates, but chlorides are also acceptable). These solutions should be standardized by titrimetric techniques such as those described by Schwarzenbach.[1] Alternatively, if a very pure form of the metal or one of its salts is available (as a primary standard) the appropriate amount may be weighed out accurately and transferred without loss to make up the standard solution. Easily hydrolyzed metal salts will require excess acid in the standard solution to prevent partial precipitation. The amount of excess acid added must be accurately known.

The supporting electrolyte must be exceedingly pure.

Because of the relatively large amount employed, and the relatively low concentration of reactants, any metal impurity at all would adversely affect the reaction being studied.

Both standard base and standard acid (e.g., KOH and HNO_3) should be made up to the same concentration as the supporting electrolyte, and should supply the same positive and negative ions as those in the supporting electrolyte. Carbon dioxide-free KOH or NaOH analytical concentrates are available from commercial sources and should be made up and standardized against potassium hydrogen phthalate and checked for CO_2 contamination. They are convenient time-savers.

4.2 The Reaction Mixture

The reaction solution is made up in a cell having features similar to that indicated in Figure 4.1. The temperature

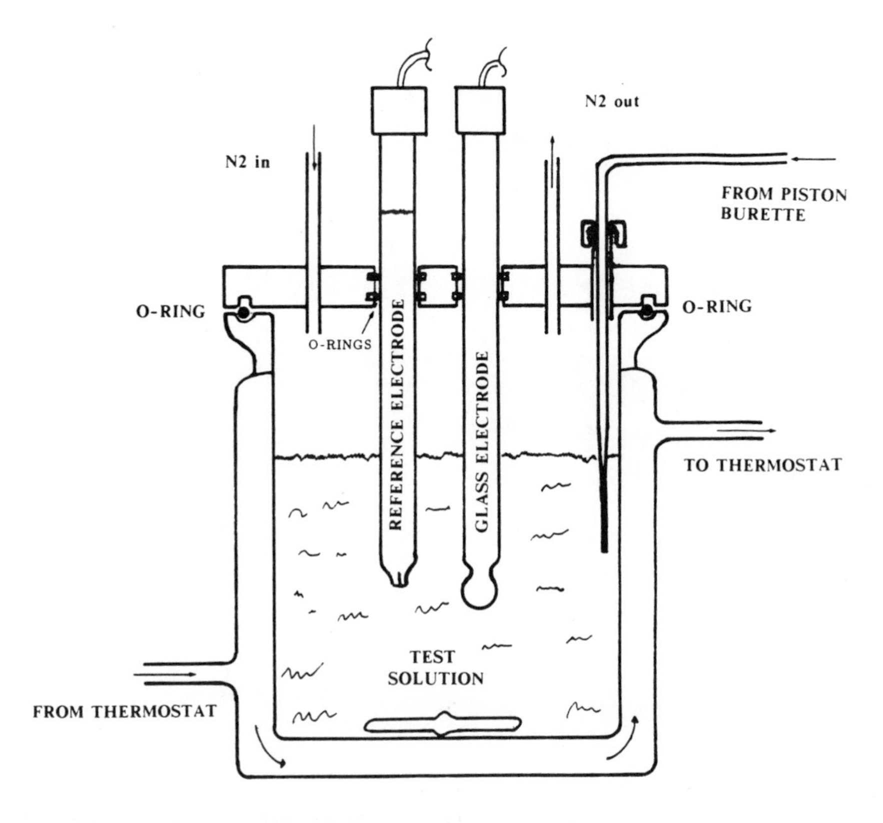

Figure 4.1. Cell for Potentiometric Measurements

is controlled by circulation of thermostated water through the jacket. The solution must be completely sealed from the atmosphere through the use of O-rings. The cap may be machined plastic or stainless steel with grooves for the O-ring seals. The inert gas must be purified to remove CO_2 and O_2 and humidified with a 0.100 M salt solution (preferably the same as the supporting electrolyte). The standard base (or acid) is added through a capillary tip beneath the surface of the solution, and is measured by a piston-type burette capable of reading volumes down to 0.01 mL or better. There should be a sufficient number of openings in the cap to take care of the electrodes and all materials to be introduced into the reaction mixture. An additional opening (not shown) is advisable for the adding of standard acid to the solution, when needed. The capacity of the cell which is recommended is about 70-80 mL, so that a volume of 50.00 mL for the experimental solution can be easily accommodated.

The reaction solution can be made up by adding precisely-measured volumes of ligand solution, standard acid if needed, metal ion solution when appropriate, and sufficient solid supporting electrolyte to provide the ionic strength desired (usually 0.100 M), allowance being made, if necessary, for the electrolyte, metal salt, and ligand contributions to the ionic strength, through the use of the ionic strength formula $\mu = 1/2 \Sigma m_i z_i^2$. Doubly-distilled water may then be added to make up the volume exactly to the pre-determined amount (50.00 mL recommended). If general practice is followed, the final solution has a low p[H], and contains the acid form of the ligand plus any excess acid, the concentration of which should be known. Carbonate-free water may be prepared by the use of commercial ion-exchange columns.

4.3 Calibration of the Potentiometric Apparatus

Prior to making measurements on the experimental solutions, it is necessary to calibrate the pH meter and electrode system in terms of p[H]. The method usually employed by the author and his coworkers is to prepare a blank solution as indicated above with no metal ions or ligands, but with the supporting electrolyte at 0.100 M. After temperature control has been achieved and the inert gas atmosphere has been established (probably not essential for calibration) and the stirrer has been started, sufficient 0.1000 M standard acid is added to bring the meter reading (which is probably not far off from the true p[H]) to about 2.5. Increments of standard base solution are then added, up to a meter reading of about 11.5. About ten or more points are needed, about equally divided between the acid and basic region and spaced two or three p[H] units away from the neutral region (ca. p[H] 7). A plot of meter reading _vs_. p[H] may be useful as an approximate calibration curve. However, the behavior of the electrodes, the pH meter, and the acid-base stoichiometry are all checked by the recomputation of the most acid points and the most basic points from the known composition of the solution. Thus p[H] = X in acid and p[H] = pK_w-X in basic solutions, where X = $-\log_{10}(|mmol_{acid} - v_{KOH}M_{KOH}|/(V_0 + v_{KOH}))$, $mmol_{acid}$ is the initial millimoles of strong acid present, v_{KOH} is the mL of standard KOH of molarity M_{KOH} added and $pK_w = -\log([H^+][OH^-])$ at the ionic strength employed.

The usual experience is that the difference between observed p[H] and $p[H]_{calc}$ in the acid region varies by <.002 up to as much as 0.004 units through values of approximately 3.5, and that the basic region differs by p[H] units of <.002 up to 0.01 - 0.015 p[H] units. It is important to avoid points close to the inflection (steep rise of p[H]) during the calibration. The largest part of the deviations observed in the basic region lie in the inevitable presence of ca. 1% or more of carbonate in these dilute calibration solutions. No corrections are usually necessary, since the readings on the acid side and the basic side are then determined by the buffering action of the system being measured, not by the trace carbonate present. The calibration line thus determined provides the factors (usually a constant factor) necessary to convert meter reading to p[H]. It is important that none of the standard adjustments on the p[H] meter be changed once the calibration has been completed. For one-point (acid) calibrations it is necessary to keep the concentration of strong acid low in order to avoid errors stemming from junction potential differences. If strong acid conditions were employed for calibration, the pH meter would be under conditions such that the H^+ ion carries a disproportionately large fraction of the current relative to the K^+ ion which constitutes the supporting electrolyte. Readings at higher p[H] values would involve conditions where H^+ is present in trace concentration and therefore nearly the entire current is carried by the K^+ ion. It is also noted that when calibrations are performed by titrating strong acid *vs*. strong base, and the amount of strong acid is large relative to the background electrolyte, there is a constant drop in the calculated ionic strength of the test solution due to the conversion of H^+ ion to water. Thus there would be insufficient ionic strength control in such a calibration.

With the new microprocessor controlled pH meters it may be more convenient to further reslope the pH meter so that the non-Nernstian response of the system may be compensated for, taking care that this compensation does not include effects of carbon dioxide contamination.

Some workers prefer to calibrate with standard buffers for which the protonation constants at the ionic strength employed are very accurately known. In order to accomplish this, at least two weak acids, which have buffer action in the acidic and basic regions, should be employed (e.g., acetic acid and potassium dihydrogen phosphate). Although it is sufficient to measure two points, one p[H] for each half-neutralized buffer, a better assessment of the solution p[H] is possible by analysis of the incremental titration of each of the buffer acids with standard base. This procedure has the added advantage of detecting the presence or absence of systematic errors which would show up as a monotonically-varying drift in the differences between measured and theoretical p[H] as titration proceeds.

Other workers prefer to bypass the pH function of the pH meter and take millivolt readings directly using the Nernst equation to convert to the appropriate hydrogen concentration function in the course of the calculations. Thus the calibration then consists of determining an E^o for the system (and possibly an effective slope).

4.4 The Experimental Runs

The first experimental run is designed to determine the protonation constants of the ligand. To the gently stirred acid solution of the ligand prepared as described above, standard base is added in sufficiently small increments to provide 50 or more (> 10 points/a value where "a" is moles base/mole ligand) experimental points for each run. Equilibrium conditions, determined by a constant meter reading falling within an interval of less than ±0.002 p[H] units must be obtained for each experimental point before proceeding with the next step. For most systems (except where the ligand is undergoing slow reactions such as isomerization, hydration or dehydration, enolization, etc.) protonation and deprotonation of the ligand is rapid and complete in the time required for mixing. The p[H] profile thus obtained for the ligand alone is used to calculate the protonation constants of the ligand by the method described below.

One or more p[H] profiles as needed are then measured for systems containing one or more molar ratios of metal to ligand. When there is an approximate match between the number of effective donor groups of the ligand and the coordination number (in solution) of the metal ion, a 1:1 molar ratio is usually sufficient. Additional ratios employed should be designed to match the expected stoichiometric ratios of complex formation between the metal ion and the ligand. The p[H] profiles thus obtained are then employed to calculate the stability constants of the complexes formed in solution, as described below using some real data.

For complex systems such as mixed ligand systems involving simultaneous combination of a metal ion with two different ligands, one must take into account the simultaneous formation of complexes in which more than one individual ligand combines with the metal ion. Thus the p[H] profiles needed to cover all possibilities are: 1, the first ligand alone, the second ligand alone; 2, the metal ion with the first ligand at more than one molar ratio (1:1, 1:2, 1:3, etc., as needed); 3, a similar set for the metal ion with the second ligand; and finally 4, the metal ion with both ligands at molar ratios that correspond to the stoichiometries of the mixed ligand complex(es) that might reasonably be expected.

4.5 Computation of Stability Constants

With the p[H] profiles described above, it is now possible to proceed with computer calculations of the stability constants. The method currently employed in the authors' research is described.

In potentiometric work, the variable measured is -log $[H^+]$, and it is therefore considered logical to carry out the calculations with an algorithm which calculates p[H] directly and minimizes the sum of the weighted squares of -log $[H^+]$ residuals. This is the feature which distinguishes programs BEST[3] and PKAS[4] from most of the programs available for the calculation of equilibrium constants. Program BEST has evolved into a very useful and friendly interactive program which, although basically designed to solve for the set of equilibrium constants corresponding to the model selected, has features which makes it possible to explore all aspects and variations of the model.

The basic algorithm in BEST can be stated in terms of equation (4.1)

$$T_i = \sum_{j=1}^{NS} e_{ij} \beta_j \prod_{k=1}^{i} [C_k]^{e_{ij}} \quad (4.1)$$

which is a statement of the mass balance of the i-th component in terms of the j-th species summed over all species present, NS. Each species concentration consists of a product of the overall stability constant and individual component concentrations $[C_k]$ raised to the power of the stoichiometric coefficient e_{ij}. For example, the EDTA-Cu system is considered as consisting of three components: $EDTA^{4-}$ (L), Cu^{2+} (M), and H^+. The species possible are $EDTA^{4-}$, $HEDTA^{3-}$, H_2EDTA^{2-}, H_3EDTA^-, H_4EDTA, $CuEDTA^{2-}$, $CuHEDTA^-$, H^+, and OH^-. There would be three mass constraints in terms of total ligand, total metal ion, and total initial hydrogen concentration: T_L, T_M, T_H, respectively.

$$T_L = [L^{4-}] + [HL^{3-}] + [H_2L^{2-}] + [H_3L^-] + [H_4L] + [CuL^{2-}] + [CuHL^-] \quad (4.2)$$

$$T_M = [Cu^{2+}] + [CuL^{2-}] + [CuHL^-] \quad (4.3)$$

$$T_H = [HL^{3-}] + 2[H_2L^{2-}] + 3[H_3L^-] + 4[H_4L] + [CuHL^-] + [BASE] + [H^+] - [OH^-] \quad (4.4)$$

In (4.4) T_H represents the amount H initially present and [BASE] that which has been removed by the added titrant (e.g., KOH). As implied in (4.1), the internal computer representation of (4.2)-(4.4) is set up in terms of β's, and the concentrations of the individual species, as expressed by (4.5)-(4.7)

$$T_M = [M^{2+}] + \beta_{ML}[M^{2+}][L^{4-}] + \beta_{MHL}[M^{2+}][H^+][L^{4-}] \quad (4.5)$$

$$T_L = [L^{4-}] + \beta_{HL}[H^+][L^{4-}] + \beta_{H_2L}[H^+]^2[L^{4-}] + \beta_{H_3L}[H^+]^3[L^{4-}] + \beta_{H_4L}[H^+]^4[L^{4-}] + \beta_{ML}[M^{2+}][L^{4-}] + \beta_{MHL}[M^{2+}][H^+][L^{4-}] \quad (4.6)$$

$$T_H = \beta_{HL}[H^+][L^{4-}] + 2\beta_{H_2L}[H^+]^2[L^{4-}] + 3\beta_{H_3L}[H^+]^3[L^{4-}] + 4\beta_{H_4L}[H^+]^4[L^{4-}] + [H^+] - \beta_{OH}[H^+]^{-1} + \beta_{MHL}[M^{2+}][H^+][L^{4-}] \quad (4.7)$$

The set of simultaneous equations (4.1 or 4.5-4.7) is solved for each component $[C_k]$. The value of $[C_k]$ is special when it represents the calculated concentration of H^+, which then is compared with the measured hydrogen ion concentration. This calculation process is repeated at all measured equilibrium points. In any calculation based on a p[H] profile there will be some known, previously calculated, β values as well as the unknown values to be determined. The first pass of the calculation procedure uses estimated values of the unknown constants.

Thus the use of the algorithm for computing equilibrium constants in BEST involves the following sequence: 1, start with a set of known and estimated overall stability constants (β's) and compute $[H^+]$ at all equilibrium points; 2, compute the weighted sum of the squares of the deviations in p[H] as in (4.8)

$$U = \Sigma w(p[H]_{obs} - p[H]_{calcd})^2 \tag{4.8}$$

where $w = 1/(p[H]_{i+1} - p[H]_{i-1})^2$, a weighting factor which serves to lessen the influence of the less accurate p[H] values in the steeply sloped regions of the p[H] profile on the calculation; 3, adjust the unknown stability constants and repeat the calculations until no further minimization of U (i.e., the sigma fit has been minimized) can be obtained, thus providing the final calculated β values. The standard deviation in pH units is obtained by the use of equation (4.9)

$$\sigma_{fit} = (U/N)^{1/2} \tag{4.9}$$

where $N = \Sigma w$.

The program PKAS is a special case of the more general algorithm found in BEST. In fact both programs can be used to calculate the protonation constants of a multidentate ligand. However since only two components (L and H) are present when metal ions are absent, much simpler generalized equations become available without the need to solve for more than two simultaneous equations (T_L and T_H). In terms of CPU time, the algorithm in PKAS is quite fast for this special case since far fewer algebraic manipulations are made relative to the generalized treatment in BEST. The distinguishing differences in the refinement of the protonation constants by the use of PKAS rather than BEST include the initial advantage of selecting reasonably close approximations of the protonation constants from the region of the p[H] profile with which the protonation constant in question is most closely involved, and the ability to use the corresponding ranges of "a" value as a basis for adjustment to the corresponding log K_i^H values based on the magnitude and sign of the differences between the observed and computed average p[H]'s. This greatly speeds up the calculation.

References

1. Schwarzenbach, G.; Flaschka, H. "Complexometric Titrations"; Methuen: London, 1969.
2. Lewis, G. H.; Randall, M. J. Am. Chem. Soc., **1921**, 43, 1112.
3. Motekaitis, R. J.; Martell, A. E. Can. J. Chem., **1982**, 60, 2403.
4. Motekaitis, R. J.; Martell, A. E. Can. J. Chem., **1982**, 60, 168.

5

COMMON ERRORS AND THEIR ELIMINATION OR MINIMIZATION

There are many sources of errors in the determination and computation of stability constants, and one of these can be the improper use of computers. In the case of PKAS and BEST, the computer will maximize the fit by minimizing the discrepancy between the measured and calculated p[H] values. With most other programs the maximized fit would be between, for example, the experimental and calculated moles of acid or base added to the system. It is apparent therefore, that the errors will arise from two principal sources, the quality of the measured data, and the model (i.e., individual species assumed to be present) provided to the program.

A word of practical advice to novices in particular is to first "get your feet wet" by obtaining several commercially pure ligands such as citric acid or NTA and simply try them out, do computations, and gain an appreciation of and a feel for the method, the equipment, and the program.

5.1 Measurement Errors

Inaccurate data may result from several sources, such as incorrect calibration of the p[H] meter-electrode system, faulty or drifting measurements, problems with the behavior of the glass and reference electrodes, impure or otherwise incorrectly characterized ligand, improper standardization of acid, base and/or metal salt solutions, and unwanted or unperceived reactions or phase separations occurring during the time of the measurements which alter the concentrations of the components (M^{n+}, L, H^+, OH^-, etc.). Although seldom a problem because they can be easily recognized, there are two other common and sometimes overlooked sources of erroneous data: an electronically faulty meter and recording errors.

5.2 Calibration and Electrode Care

It is most convenient to calibrate the potentiometric apparatus in terms of p[H], as described above (Section 4.3), by means of a calibration "curve" (usually a straight line), between $[H^+]$ values of ca. 10^{-3} and 10^{-11} M. Alternatively, millivolt readings may be calibrated in terms of p[H]. It is not considered advisable to employ "pH standards", obtained from chemical supply companies because these buffer systems usually do not indicate either the method of determination or the activity coefficients employed and are guaranteed only to approximately 0.01-0.02 "pH" units. The experimentalist may, however, make up other p[H] standards, such as acetate (CH_3COO^-, CH_3COOH) or phosphate (HPO_4^{2-}, $H_2PO_4^-$) buffers at controlled ionic strength, because activity coefficients have been determined for these systems, and p[H] can be calculated accurately to three decimal numbers.

When not in use, the glass electrode should be stored by immersion in the supporting electrolyte containing ca. 0.001 M H^+. Occasionally, the glass electrode shows erratic behavior after being in contact with a solution of a heavy metal ion. In most cases it may be restored to normal function by being immersed in 0.10 M HCl for an indefinite period or by following manufacturer's recommendations.

Malfunction of the reference electrode may result from crystallization of KCl in the opening that provides the liquid junction to the experimental solution, thus preventing normal conductivity between the electrodes. Such obstructions should be removed. Alternatively, use of pressure in the cell may cause changes in the measured potential if some of the experimental solution is forced into the reference electrode compartment, thus changing the liquid junction potential in a non-reproducible way. A reference electrode should be kept filled and should never be stored (nor used) deeply immersed, to prevent backflow of experimental solution into the electrode compartment.

There are several special extenuating circumstances when the commercially obtained saturated KCl (or 3M KCl) reference electrode becomes particularly unsuitable, resulting in the formation of a solid plug in the small orifice (liquid junction) with consequential serious impairment in its performance and increasing drift. These difficulties are almost certain to arise when $NaClO_4$ is employed as the background electrolyte or when organic-rich mixed solvents are used. In the former case, the insoluble $KClO_4$ is the problem, while in the latter the KCl in the orifice may be salted out by the organic solvent present.

While the employment of a secondary salt bridge could solve the ClO_4^--induced problem, it is usually far more convenient to replace the electrode solution with NaCl solution. For mixed solvent studies it is advisable to cut back on the concentration of the solution of KCl in the reference electrode while keeping in mind that the accessible p[H] range of measurement is decreased. In practice, in our laboratories, we make a correction table based on observed vs. calculated $p[H^+]$ using standard acid, standard base, and 0.100 M KCl-filled electrodes for mixed solvent studies. It takes at least two days for the reference electrode to stabilize after its composition has been altered.

5.3 Reagents

Inaccurate standardization of the reagents could be the source of considerable error, and must be avoided. Thus a small error in the concentration of a tetraprotonated or tetrabasic ligand (H_4L) would produce a four-fold error in the concentration of the available hydrogen ion, T_H. While this may not dramatically affect the equilibrium constants for the major species present, it could result in disproportionate errors in the constants calculated for minor solution species. Ligands are generally employed in the form of one of their acid species; when a ligand contains many protonated functional groups it sometimes crystallizes with a non-stoichiometric amount of excess acid, which must be determined. Similarly, stock solutions of many of the salts of the less basic metal ions must contain excess acid to prevent partial hydrolysis and precipitation. One standard method of determining excess acid is the use of the Gran's plot.[1]

The principle behind Gran's method for strong acids is that it is assumed that free hydrogen ion concentration $[H^+]$ is reduced in proportion to the amount of standard alkali hydroxide added. The function plotted is $(V_o + v_{KOH})10^{-DR}$ against v_{KOH}, where V_o is the initial volume; v_{KOH} is the volume of standard alkali, and DR is the pH-meter dial reading (or digital output reading) on the log scale. Of course, DR is equal to p[H] plus some offset or calibration factor. Thus the factor 10^{-DR} is exactly proportional to $[H^+]$ and the overall term $(V_o + v_{KOH})10^{-DR}$ is proportional to the moles of hydrogen ions present. The intercept on the v_{KOH} axis then represents an exact neutralization of the strong acid present, even if a hydrolysis reaction or other weak acid equilibrium takes place at moderately acid pH values.

When the metal salt employed is sufficiently pure (e.g., ACS reagent grade or equivalent) the excess acid employed may be predetermined by the addition of a measured volume of standard acid to the standard metal ion solution. In the case of the ligand, the unknown quantity of acid (H^+) may be determined by treating it as an additional unknown in the computation. This can be done only when the ligand is otherwise of high purity. Similarly, if a ligand is impure, but the impurities are known to be inert but in undetermined amount, the "molecular weight" of the ligand and impurities may be determined by allowing the total ligand molarity to float as an unknown. A recent example is the initial potentiometric study of BISTREN,[2] which was available as the free base in pure form but as a syrup containing an unknown amount of water and carbonate. Its "molecular weight" (i.e., the molarity of a stock solution) was treated as an unknown. The absorbed carbon dioxide present as an impurity was not inert, but was eliminated by passing purified nitrogen through the solution containing added excess acid for an extended period of time before beginning the experimental determination.

There are several other precautions, in addition to those described above, which should be observed:- maintenance of an inert atmosphere, careful temperature control, adjustment of the ionic strength with a supporting electrolyte, etc. It is important that the supporting electrolyte be truly inert. While KNO_3 is satisfactory for most systems, KCl and $NaClO_4$ are frequently employed. However, Na^+ and K^+ are complexed by some ligands (e.g. crown ethers, cryptands), while many chloro complexes are known, as well as some nitrate complexes (e.g., of protonated cryptands). To avoid complications from such weakly complexing cations and ions, supporting electrolytes such as tetramethylammonium perchlorate are suggested.

Temperature control is necessary for several reasons. Electrode response for a given concentration of $[H^+]$ is a function of temperature. Its variation gives rise to a cumulative effect including all of the additive potentials: liquid-glass, liquid-liquid, solid-liquid, as well as concentration potentials. In addition, depending on the value of $\Delta H°$ for a given equilibrium, there is usually a significant effect on the chemical equilibrium itself, i.e., K_{eq} is a function of temperature.

5.4 Equilibrium Measurements

Before an experimental point (p[H]) can be measured, sufficient time must be allowed for establishment of equilibrium. There are two general requirements: that electrode systems be given sufficient time to adjust to the new conditions, and that limitations imposed by the reaction rates of the species undergoing complex formation and dissociation be addressed. The electrode response requirement is usually complete in about a half minute if the new p[H] reading is within <0.1 of the previous value, and if the stirring is sufficiently rapid to efficiently mix the contents. In the extreme case, electrode response time may be longer than 10 minutes when the p[H] change is more than one or two units. It has been the experience of these authors and their coworkers that the reference electrode becomes unstable and sensitive to stirrer speed when the solid KCl present in the reference (saturated calomel) electrode is allowed to plug the immersed orifice. An electrode in this condition must be serviced (i.e., the channel must be cleared) immediately.

In the absence of slow chemical reactions, slow response time, especially above p[H] 7, can be sometimes attributed to a glass electrode which has become worn out, or has been used repeatedly with polyvalent metal ions. Usually, the worn out electrode continues its slow drift for hours, approximately 0.001 pH units/min, while the other stops drifting. The worn out electrode cannot be rejuvenated with acid treatment, while the poisoned one can be revived to as good as new following manufacturer's recommendations.

The rate of chemical equilibration depends on the system studied. In cases where equilibrium is slow, one way to ascertain with reasonable certainty whether equilibrium has been achieved is to perform the experiment a second time allowing longer equilibration time. If, upon addition of titrant, equilibrium seems to stabilize within about 10 minutes, an additional 10 minutes should be allowed, to determine if all drifting has ceased. If the initial equilibrium appears to take approximately 2 hours or more, then it becomes impractical to obtain quality data over the lifetime of the experiment (>40 hours) because then it cannot be guaranteed that the electrode system will remain calibrated for such a long period of time. Other possible problems that may develop with slow potentiometric equilibrium experiments involve titrant diffusion from the capillary burette tip (burette tip may be raised above solution level during long equilibrations), significant increase in ionic strength resulting from contamination of the test solution by the electrolyte in the reference electrode, possible unanticipated side reactions or decomposition of the ligand, and separation of a solid phase. Two remedies are possible: a slowly equilibrating system may sometimes be studied either through an incremental method wherein several data points are obtained with as much certainty as possible before the solution is discarded and replaced by a new solution; or through the use of a batch method wherein individual sealed thermostatted containers representing successive data points are made up and stored for the required period of time, and are then checked and rechecked to ascertain the equilibrium p[H] value.

Kinetically inert or sluggish equilibrium systems must always be treated with special precautions. The slowly equilibrating systems can vary in their behavior in a wide variety of ways, so that each system must be given individual treatment designed to maximize the accuracy of the measured data.

Automatic titrators or data collectors are primarily designed for analytical titration purposes, which means that they concentrate on end-point accuracy. Such devices frequently possess another mode which can be preset for definite time increments between additions of titrant. In light of the above discussion, it appears that commercial automatic titrators of this type would then be useful mainly for the measurement of rather labile equilibria. Perhaps a more useful device would be a microprocessor-controlled burette programmed to make decisions, similar to those described above, concerning whether equilibrium has been achieved.[3] The construction and use of equipment of this type has recently been reported.[4] In this particular case, the authors claim that a maximum of four minutes between points is sufficient. It is believed that while this protocol may be one of convenience and that it may indeed be sufficient for most labile systems, it is exceedingly difficult to program the way a system behaves or an investigator thinks. Therefore there may be pitfalls resulting from systematic errors which could easily go undetected even with what one would believe to be a conservative timing program. For example, consider a system which is very slow in reaching equilibrium and is being measured by a microprocessor which samples points every second, and is programmed to finalize the measurement when 60 successive p[H] readings are within ±0.0005 of the mean. If the equilibration rate lowers the p[H] by 0.001 unit/2 minutes, then in one hour the p[H] reading would have been about 0.030 units lower than the one actually recorded by this microprocessor. In three hours, this point would be some 0.050 p[H] units lower. Because the above algorithm would take data at similar time intervals, the recorded curve would be higher (in p[H]) than the actual equilibrium curve but would otherwise closely resemble it in appearance. The net result of this error would be a lower stability constant; nevertheless the internal consistency or fit of the calculated to the experimental p[H] profile would probably be very good (i.e., the apparent error would be small).

5.5 **Calculations**

Internal consistency and a good sigma fit are rather easy to achieve because the curve fitting algorithm is set up to do just that. Internal accuracy means that reproducibility is possible, and that the equipment is in apparently good shape, but does not guarantee absolute accuracy. On the other hand, poor internal accuracy or precision is a sign of either poor data, a poor model, or both. If the data are not suspect, there must be something wrong with the model. A poor or incorrect model is either incomplete or an incorrect statement of species present. Therefore, once it is ascertained that the data are accurate, the model may be modified one item at a time in accordance with sound principles of coordination chemistry until a satisfactory refinement (low σ_{fit} to experimental data) is obtained.

This approach is widely used but has been criticized as mathematically arbitrary by Vacca et al.,[5] who advocate the use of rigorous hypothesis testing to determine if one of the proposed models is to be preferred. In turn, Jaskolski and Lomozik[6] emphasized similar concerns. The problem of ascertaining the correctness of the model is expressed in many current publications. Meloun and coworkers[7] focus in on equilibrium model determination through the use of multiple electronic spectra, while Fournaise and Petitfaux[8] consider the factors involved in reducing errors in potentiometric model selection.

Some of this detailed consideration of model selection,

however, is unjustified, since the models employed involve dubious cases requiring choices between minor species. Since all least squares algorithm minimize the fit between calculated and observed data, it is inevitable that all experimental errors end up ultimately refined into the β values themselves. It is in cases where the errors are particularly large that there is temptation to invent species to cover cases for which a satisfactory fit cannot be achieved with a reasonable and sensible model.

Finally, it is important to take a good, hard look at the resulting stability constants and compare the nature and magnitudes of the concentration quotients with those of previously reported systems. It is instructive to compare the constants obtained with those of the same metal ion with similar ligands, or those of the same or similar ligands with other analogous metal ions. For example, if the p[H] profile of alanine with Cu(II) is measured and log K_{ML} calculated as 12.65 with a $\sigma_{fit} = \pm 0.010$, but a comparison with the literature reveals that the formation of the Cu(II) complex of a similar ligand, glycine, is characterized by a log K_{ML} of 8.15, the alanine result should be considered doubtful, and should be reinvestigated.

How could it be possible to obtain a good sigma fit for the example described above when the magnitude of its constant is so far off? In this case, as in many other cases, the Cu(II) ion concentration in the experimental solution was extremely low because most of the metal ion was bound to the ligand even at the lowest p[H] values. Thus the starting p[H] already reflects the complete or nearly complete displacement of the active protons of the ligand as the result of complete or nearly complete formation of CuL^-. This condition is satisfied by the true formation constant (log K_{ML} = 8.13) as well as by higher formation constants, including log K_{ML} = 12.65. A simple check of the concentrations of the species present at the lowest p[H] would indicate whether there is any appreciable concentration of free metal ion. If not, then the corresponding data cannot be used for the determination of the desired formation constant. When at equilibrium if the reaction is either substantially incomplete at the end or complete at the beginning of the titration, the data obtained cannot be used for determination of stability constants without incurring large errors in the calculated values. Significant concentrations of all of the equilibrating species must be present simultaneously if an accurate formation constant is to be computed.

For example, leucine possesses a very weak tendency to complex with calcium(II) ion, and therefore does not form its CaL^+ complex until p[H] > 8 is reached. Because the equilibrium constant log K_{ML} is only 2.40, under the usual experimental solutions (ca. 10^{-3} M) the complex is less than 20% formed. Since the measured pH depends largely on the protonation constant, but not so much on the displacement by Ca(II), a small error (ca. 1% too large) in the ligand concentration would result in a log K_{ML} of 2.26 or 0.14 log units too small: about a 28% error.

Thus it is seen that absolute accuracy, even though it is sometimes difficult to obtain, always requires a carefully calibrated pH meter-electrode system, a low σ_{fit}, the assurance that the stoichiometric variables are accurate, and the presence of a substantial proportion of each species in the equilibrium expression under the conditions of the experiment. An effective test of whether these last conditions are met is to plot the distribution curves of the system (e.g., Figure 2.1) by employing the β values under consideration, to demonstrate that all reacting species are present at reasonable (i.e., measurable) concentrations.

As a built-in convenience program BEST has simple commands for displaying the species compositions at any desired p[H] values at any point in the refinement process. In the above example, it would have become evident that with a log K_{ML} of either 12.65 or 8.13, the free $[Cu^{2+}]$ is ca. 0.00% and [CuL] is ca. 100% formed throughout the titration pH range. At this point it would have been realized that the computation would have been impossible throughout the p[H] range of the measurements.

Data on new ligands possessing unique structures or unique metal-ligand combinations cannot be compared to literature values. However, the results obtained for such systems should always be carefully considered to make sure that the principles and guidelines of coordination chemistry are satisfied, and that the requirements for equilibrium constant calculations described above are met.

5.6 Selection of the Model

Some reports have raised the question of uniqueness of the model. Can several models fit the same data? In our experience, such ambiguity may be possible only if the system is entirely unpredictable, or very complex, or no previous analogies exist, or if a large number of minor species are present, or if there are relatively large errors in the data. Major species are no problem and their stability constants are readily determined with reasonable certainty. It is with the minor species that one may encounter trouble with either uniqueness or uncertainty. This is both a problem and a non-problem. It becomes a problem if there is no alternate way to determine the presence or absence of a given species. The most common alternate ways are aqueous NMR (in D_2O solution), aqueous IR, and electronic spectroscopy. Also available is the possibility of performing additional potentiometric determinations involving ratios of components chosen so as to maximize the presence of the minor species in question. Sometimes the best one can do is to estimate the formation constant of such minor species.

Of course there is another kind of non-uniqueness of the model. For example data taken for a solution containing 0.2 millimoles of a monobasic ligand can just as well be fitted to a model containing 0.1 millimole of a dibasic acid and two protonation constants, or even 0.05 millimoles of a tetrabasic acid with four protonation constants. Such confusion is not likely in practice, since in general one knows both the molecular weight and the number and identity of the acidic or basic groups present.

An effective way to be sure of the nature of the species present, when doubt exists about its degree of aggregation, is to measure the equilibria at several concentration levels of the species in question. The shift of equilibria observed will vary with the number of particles in equilibrium with the species being measured, thus making it possible to select the correct one. If the value of the calculated constant is found to be independent of the concentration of the components, then it is safe to assume that only mononuclear complexes form. Otherwise the model must include one or more polymeric forms.

References

1. Gran, G. Analyst, **1952**, 77, 661.
2. Motekaitis, R. J.; Martell, A. E.; Lehn, J. M.; Watanabe, E. I. Inorg. Chem., **1982**, 21, 4253.
3. Harris, W. R., private communication.
4. Gampp, H.; Maeder, M.; Zuberbuhler, A. D.; Kaden, T. A. Talanta, **1980**, 27, 513.
5. Vacca, A.; Sabatini, A.; Gristina, M. A. Coord. Chem. Revs., **1972**, 8, 44.
6. Jaskolski, M.; Lomozik, L. Talanta, **1982**, 32, 511.

7. Meloun, M.; Javurke, M; Havel, J. Talanta, **1986**, 33 (6), 513.
8. Fournaise, R.; Petitfaux, C. Talanta, **1986**, 33 (6), 499.

EXAMPLES OF STABILITY CONSTANT DETERMINATIONS

6.1 Iminodiacetic Acid (IDA)

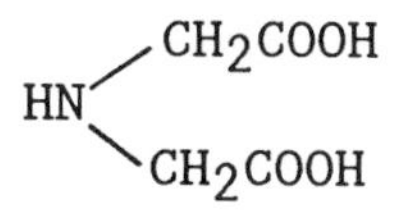

Iminodiacetic acid (IDA) is a well-known terdentate aminodicarboxylic acid which is capable of coordinating with a variety of metal ions to form complexes having 1:1 and 1:2 molar ratios of metal to ligand. As an example of a straightforward potentiometric determination of stability constants, the detailed step-by-step procedure for the determination of the stability constants of IDA with Pb(II) is now described. The purpose of this example is to provide a typical description of both the experimental and computational methods used in such potentiometric determinations.

The following experimental materials and solutions were employed:

- Double-distilled or deionized water
- 0.1024 M CO_2-free KOH[a]
- 1.00 M KCl solution[b]
- 0.02305 M $Pb(NO_3)_2$ solution[c]
- 0.01234 M IDA[d]
- 0.1352 M HCl[e]

---

[a] Standardized with KHP (potassium biphthalate). [b] Prepared by weighing out the purest KCl available. [c] Prepared by dissolving ca. 3.6 g $Pb(NO_3)_2$ in 500 mL water and standardizing with Dowex 50W X-8 Ion Exchange Resin in H^+ form. (20.00 mL of lead(II) nitrate solution required 9.28 mL of 0.09934 M KOH). [d] Prepared (from commercial disodium salt by acidification with hydrochloric acid and recrystallization from aqueous ethanol) by weighing 0.821 g of the acid form into 500.0 mL water. [e] Standardized with KOH by the use of Gran's method[1] (Section 5.3).

6.2 Procedure for IDA

Calibration. A 5.000 mL quantity of 0.1352 M HCl solution was added to the equilibrium reaction cell (Figure 4.1) containing 5.000 mL of 1.000 M KCl in 45.00 mL water. The cell was capped (with thermostat setting at 25.00°C and inert gas flow), equilibrated for about a half hour, and the pH meter was set to read p[H] 1.910. This calculation is based on the equation:

$$p[H] = -\log_{10} [H^+]$$

$$\text{where } [H^+] = (0.1352 \text{ M HCl})(5.000 \text{ mL HCl}/55.00 \text{ mL}).$$

In this example, the quantity of HCl used is a little greater than usually recommended, resulting in some 13% contribution toward the 0.1 M ionic strength by HCl alone. Accurate calibration requires <10% contribution by the acid added. This quantity of acid is borderline between acceptable and too much. It is a valid amount nevertheless, since further tests show that the deviations are not significant.

Checking the Electrode System. 1.000 mL quantities of 0.1024 M KOH solution were added to the above solution while the pH meter readings were recorded (see Table 6.1). Note that these data indicate an excellent match in the acid region, but the lack of agreement in the basic region points to the presence of some carbonate in the experimental solution. In order to assess the degree of carbonate contamination the method of Gran[1] is used here. Values of ϕ are plotted against the volume of KOH used as shown in Figure 6.1,

$$\phi = (V_0 + V_{KOH}) \times 10^{\pm R}$$

where V_0 = initial volume

V_{KOH} = volume of KOH added

R = reading on pH meter

"-" = used in acid region

"+" = used in alkaline region

Table 6.1. Strong Acid vs. Strong Base

mL KOH	$p[H]_{obs}$	$p[H]_{calc}$	ϕ
0.000	1.909	1.910	0.6782
1.000	1.988	1.989	0.5757
2.000	2.080	2.082	0.4741
3.000	2.195	2.196	0.3702
4.000	2.345	2.345	0.2666
5.000	2.559	2.563	0.1656
6.000	2.996	2.994	0.0616
7.000	10.163	10.595	0.090E-13
8.000	11.031	11.136	0.677E-13
9.000	11.305	11.364	1.292E-13
10.000	11.467	11.508	1.905E-13

$\phi = Vx10^{-pH_{obs}}$ on the acid side and $\phi = Vx10^{pH_{obs}}$ on the basic side. V = total volume of solution.

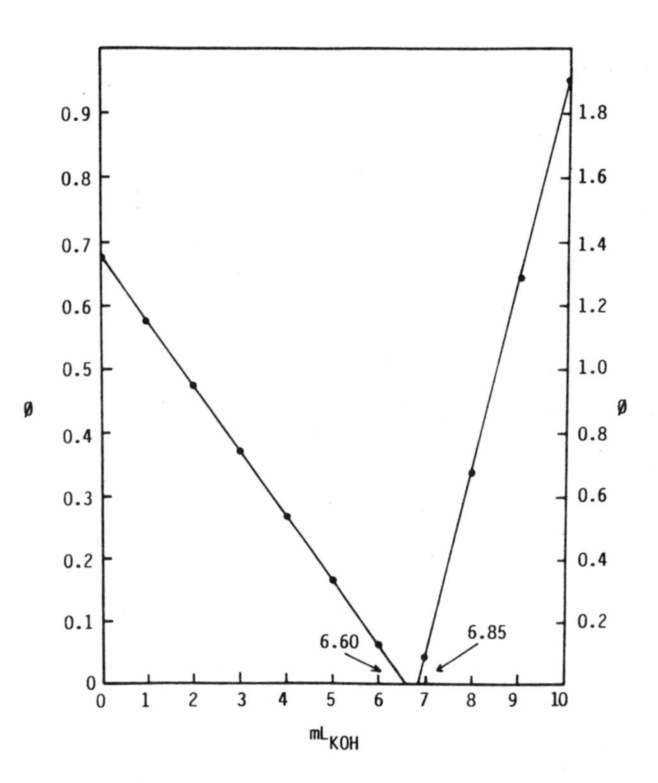

Figure 6.1. Gran's Plot of ϕ vs. ml_{KOH} (see footnote to Table 6.1 for definition and values of ϕ)

yielding two straight lines intercepting on the volume axis. The left line ϕ is proportional to $[H^+]$) intersects at 6.60 mL KOH while the right sloping line (ϕ is proportional to $[OH^-]$) intersects at 6.85 mL KOH. The difference 0.25 mL reflects the carbonate which is about 1.8% (100 x 0.25/2/6.85) of the KOH present. Commercial "CO_2-free" ampoules often contain from 0.5 to 1.5% carbonate. Over 2% carbonate becomes troublesome and should be eliminated by preparation of a fresh solution. For example, the $p[H]_{obs}$ values in the basic region of Table 6.1 are too low and are not suitable for the accurate calibration of the basic region.

Initial Data for IDA. The cell is charged with 20.00 mL 0.01234 M IDA solution, 5.00 mL of 1.000 M KCl and 25.00 mL water. It is closed, equilibrated and slowly titrated with 0.100 mL increments of 0.1024 M KOH solution until at least enough KOH is added to neutralize the titratable acid present. (The ionic strength of the solution is nominally 0.100 M initially and remains nearly constant since it is diluted with 0.1024 M KOH.) The data obtained are listed in Table 6.2. Stability constant refinement is the process whereby the computer adjusts a parameter or several parameters with a programmed algorithm for the purpose of obtaining the best possible least-squares fit of calculated data to the observed data. The goodness of fit is measured by σ_{fit} which is a measure of the weighted sum of the squares of the $p[H]_{cal}$-$p[H]_{obs}$ differences and the object is to minimize σ_{fit} through refinement of these parameters. $\sigma_{fit} = (U/N)^{1/2}$, where U is defined in equation 4.8. Usually the parameters are log overall stability constants (β's), however with proper justification the refinement may involve T_H (usually) or even T_M. The refinement process starts with a specified set of fixed and adjustable parameters and is completed when no combination of adjustments can give rise to a lower σ_{fit}.

Table 6.2. Titration Data for Iminodiacetic Acid

mL KOH	p[H]	mL KOH	p[H]	mL KOH	p[H]
.000	2.675	1.600	3.412	3.200	9.262
.100	2.702	1.700	3.528	3.300	9.332
.200	2.728	1.800	3.862	3.400	9.403
.300	2.765	1.900	3.914	3.500	9.476
.400	2.786	2.000	4.401	3.600	9.549
.500	2.818	2.100	7.120	3.700	9.625
.600	2.852	2.200	8.012	3.800	9.703
.700	2.888	2.300	8.310	3.900	9.781
.800	2.927	2.400	8.498	4.000	9.867
.900	2.968	2.500	8.639	4.100	9.957
1.000	3.013	2.600	8.756	4.200	10.055
1.100	3.061	2.700	8.861	4.300	10.259
1.200	3.115	2.800	8.952	4.400	10.363
1.300	3.174	2.900	9.036	4.500	10.463
1.400	3.242	3.000	9.114	4.600	
1.500	3.320	3.100	9.190		

The preliminary estimates of the log protonation constants of IDA were 9.5 and 3.0, or 9.5 and 12.5 in overall (β) units. These were read directly from the p[H] profile at a-values near 1.5 and 0.5, respectively, of Figure 6.2. These values were then refined until sigma was minimized. The results are shown in Table 6.3.

The expected millimoles of titratable hydrogen ion (0.4938) is two times that of the millimoles of ligand IDA (0.2469) present, since IDA is a dibasic (H_2L type) acid. However, the results of the preliminary refinement of the overall protonation constant shown in Table 6.3 indicate that something is wrong. Indeed, recalling that the acid form of IDA was precipitated from an acidified aqueous alcohol solution of the sodium salt of the ligand, a correction on the hydrogen ion present initially in the diacid form in the solid material is necessary. This is indicated by the generally bad fit and in particular by the position of the inflection of the p[H] profile halfway through the titration.

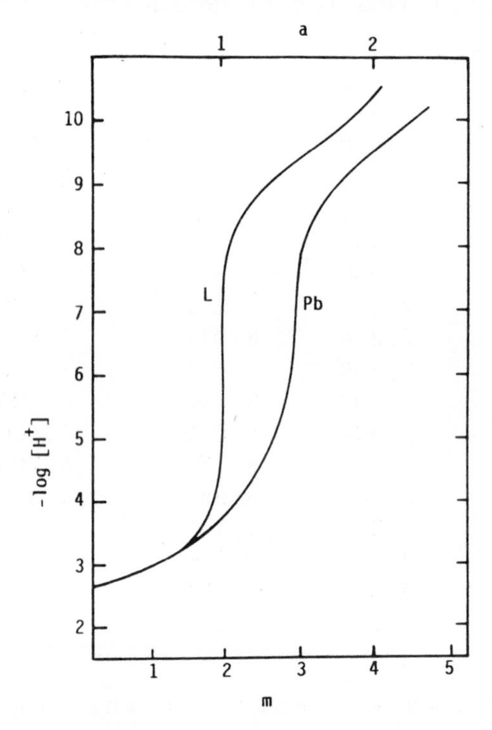

Figure 6.2. Potentiometric Equilibrium Curves for IDA (L) and 2IDA/1Pb (Pb) as a Function of Added KOH. a = moles of base added/moles of ligand present, m = moles base added/moles metal ion present

Table 6.3. Preliminary Results on IDA Protonation Calculation.

```
Program BEST vers. 05/17/88
         IDA Iminodiacetic Acid
         INITIAL VOLUME                        50.00000
         NORMALITY OF BASE                       .10240
         MILLIMOLES ACID                         .00000
         NUMBER DATA POINTS                          47
         PH CORR. INCLUDED                         .000

 COMPONENTS:
       1 IDA           .24690  MILLIMOLES
       2 H             .49380  MILLIMOLES

 SPECIES:
   LOG BETA
 1     .0000  1 IDA      0 H
 2    9.6429  1 IDA      1 H
 3   12.5109  1 IDA      2 H
 4     .0000  0 IDA      1 H
 5  -13.7800  0 IDA     -1 H

         SIGMA PH FIT=    .221098
```

Table 6.3. continued

DIFFERENCE TABLE:("DIFF"=LOGBETA(I)-LOGBETA(I-1))

I	DIFF	ERROR
1	.0000	.0000
2	9.6429	.0577
3	2.8680	.0971
4	-12.5109	.0000
5	-13.7800	.0000

	VB	A	PH	PHCALC	DIFF
1	.000	.000	2.675	2.700	-.025
2	.100	.041	2.702	2.729	-.027
3	.200	.083	2.728	2.759	-.031
4	.300	.124	2.756	2.790	-.034
5	.400	.166	2.786	2.822	-.036
6	.500	.207	2.818	2.855	-.037
7	.600	.249	2.852	2.890	-.038
8	.700	.290	2.888	2.926	-.038
9	.800	.332	2.927	2.964	-.037
10	.900	.373	2.968	3.004	-.036
11	1.000	.415	3.013	3.045	-.032
12	1.100	.456	3.061	3.089	-.028
13	1.200	.498	3.115	3.136	-.021
14	1.300	.539	3.174	3.185	-.011
15	1.400	.581	3.242	3.238	.004
16	1.500	.622	3.320	3.295	.025
17	1.600	.664	3.412	3.358	.054
18	1.700	.705	3.528	3.427	.101
19	1.800	.747	3.682	3.504	.178
20	1.900	.788	3.914	3.593	.321
21	2.000	.829	4.401	3.700	.701
22	2.100	.871	7.120	3.832	3.288
23	2.200	.912	8.012	4.012	4.000
24	2.300	.954	8.310	4.302	4.008
25	2.400	.995	8.498	5.308	3.190
26	2.500	1.037	8.639	8.219	.420
27	2.600	1.078	8.756	8.564	.192
28	2.700	1.120	8.861	8.768	.093
29	2.800	1.161	8.952	8.917	.035
30	2.900	1.203	9.036	9.038	-.002
31	3.000	1.244	9.114	9.141	-.027
32	3.100	1.286	9.190	9.232	-.042
33	3.200	1.327	9.262	9.315	-.053
34	3.300	1.369	9.332	9.393	-.061
35	3.400	1.410	9.403	9.466	-.063
36	3.500	1.452	9.476	9.537	-.061
37	3.600	1.493	9.549	9.606	-.057
38	3.700	1.535	9.625	9.673	-.048
39	3.800	1.576	9.703	9.741	-.038
40	3.900	1.617	9.781	9.808	-.027
41	4.000	1.659	9.867	9.877	-.010
42	4.100	1.700	9.957	9.947	.010
43	4.200	1.742	10.055	10.019	.036
44	4.300	1.783	10.158	10.093	.065
45	4.400	1.825	10.259	10.170	.089
46	4.500	1.866	10.363	10.249	.114
47	4.600	1.908	10.463	10.330	.133

Final Calculation of IDA Protonation Constants. The refinement of the millimoles of titratable hydrogen present was carried out in the following way. In successive steps the, quantity of titratable hydrogen presumed present was decreased stepwise concomitant with a single pass refinement of the two protonation constants at each step. This minimization with respect to millimoles of hydrogen ion was repeated until further minimization in σ_{fit} was impossible. The total titratable acid was thus reduced from the initially assumed 0.4938 mmol down to 0.4608 mmol distributed as 0.4938 mmol associated with IDA and -0.0330 mmol as missing at the beginning of the titration (note the a-value is 0.134 mol base/mol ligand at VB = 0 in Table 6.4). This seemingly awkward concept of negative millimoles is a consequence of setting a = 0 at the point in the titration where the ligand is a neutral molecule. For IDA this point corresponds to the diprotonated form, H_2L. The improvement in the σ_{fit} is quite remarkable, from the 0.221098 shown in Table 6.3 to the value of 0.005452 shown in Table 6.4.

Table 6.4. Final Results on Determination of Protonation Constants of IDA, Showing Non-Stoichiometric Amount of the millimoles of acid present.

```
Program BEST vers. 05/17/88
        IDA Iminodiacetic Acid
        INITIAL VOLUME                 50.00000
        NORMALITY OF BASE                .10240
        MILLIMOLES ACID                 -.03300
        NUMBER DATA POINTS                   47
        PH CORR. INCLUDED                  .000

 COMPONENTS:
       1 IDA          .24690  MILLIMOLES
       2 H            .49380  MILLIMOLES
 NOTE: millimoles were adjusted

 SPECIES:
   LOG BETA
 1     .0000  1 IDA      0 H
 2    9.3431  1 IDA      1 H
 3   11.8922  1 IDA      2 H
 4     .0000  0 IDA      1 H
 5  -13.7800  0 IDA     -1 H

        SIGMA PH FIT=   .005303
```

Table 6.4. continued

DIFFERENCE TABLE:("DIFF"=LOGBETA(I)-LOGBETA(I-1))

I	DIFF	ERROR
1	.0000	.0000
2	9.3431	.0147
3	2.5491	.0114
4	-11.8922	.0000
5	-13.7800	.0000

	VB	A	PH	PHCALC	DIFF
1	.000	.134	2.675	2.668	.007
2	.100	.175	2.702	2.696	.006
3	.200	.217	2.728	2.725	.003
4	.300	.258	2.756	2.755	.001
5	.400	.300	2.786	2.787	-.001
6	.500	.341	2.818	2.820	-.002
7	.600	.383	2.852	2.855	-.003
8	.700	.424	2.888	2.892	-.004
9	.800	.466	2.927	2.931	-.004
10	.900	.507	2.968	2.973	-.005
11	1.000	.548	3.013	3.018	-.005
12	1.100	.590	3.061	3.067	-.006
13	1.200	.631	3.115	3.121	-.006
14	1.300	.673	3.174	3.180	-.006
15	1.400	.714	3.242	3.246	-.004
16	1.500	.756	3.320	3.321	-.001
17	1.600	.797	3.412	3.409	.003
18	1.700	.839	3.528	3.516	.012
19	1.800	.880	3.682	3.653	.029
20	1.900	.922	3.914	3.845	.069
21	2.000	.963	4.401	4.180	.221
22	2.100	1.005	7.120	7.016	.104
23	2.200	1.046	8.012	8.024	-.012
24	2.300	1.088	8.310	8.322	-.012
25	2.400	1.129	8.498	8.510	-.012
26	2.500	1.171	8.639	8.651	-.012
27	2.600	1.212	8.756	8.768	-.012
28	2.700	1.254	8.861	8.868	-.007
29	2.800	1.295	8.952	8.958	-.006
30	2.900	1.336	9.036	9.041	-.005
31	3.000	1.378	9.114	9.118	-.004
32	3.100	1.419	9.190	9.192	-.002
33	3.200	1.461	9.262	9.264	-.002
34	3.300	1.502	9.332	9.334	-.002
35	3.400	1.544	9.403	9.404	-.001
36	3.500	1.585	9.476	9.474	.002
37	3.600	1.627	9.549	9.545	.004
38	3.700	1.668	9.625	9.618	.007
39	3.800	1.710	9.703	9.694	.009
40	3.900	1.751	9.781	9.774	.007
41	4.000	1.793	9.867	9.859	.008
42	4.100	1.834	9.957	9.950	.007
43	4.200	1.876	10.055	10.047	.008
44	4.300	1.917	10.158	10.151	.007
45	4.400	1.959	10.259	10.260	-.001
46	4.500	2.000	10.363	10.370	-.007
47	4.600	2.042	10.463	10.477	-.014

These results point to the absence of some 0.0330 millimoles of hydrogen ion, as if ca. 6.6% of the hydrogen ion were pre-neutralized in the stock solution. This is a typical example whereby the ligand is really chemically pure, but possesses a non-stoichiometric number of hydrogen ions. Because of the method of isolation and the low first pK_a, some monosodium hydrogen IDA was present with the acid (dihydrogen form) of the ligand isolated from the alcohol-water solvent mixture.

Thus at 25.00° C and ionic strength 0.100 M the log protonation constants obtained for IDA are 9.34 and 2.55. These values compare well with the critical literature data found in Martell and Smith:[2] 9.32 and 2.61. With these accurate protonation constants, the state of protonation of the ligand is known at all p[H] values and the determination of the stability constants of the lead(II)-IDA complexes is now carried out.

Lead(II)/IDA System. The cell is charged with 20.00 mL 0.01234 M IDA solution (0.2469 millimoles IDA), 5.000 mL of 0.02305 M $Pb(NO_3)_2$ (0.1152 millimoles Pb^{2+}), 5.00 mL of 1.000 M KCl and 20.00 mL water. The vessel is sealed, equilibrated, and slowly titrated with 0.1024 M KOH in 0.100 mL incremental additions until at least enough KOH is added to somewhat more than neutralize the titratable acid present. The mL vs. p[H] data (p[H] profile) obtained are listed in Table 6.5. These data are plotted in Figure 6.2.

One next examines the Pb-IDA p[H] profile while keeping in mind that there are two moles of IDA's per mole of Pb(II) (actually 2.14/1.00). The coincidence of the curves in the acid region for the ligand alone and the ligand and Pb(II) containing systems indicates that the Pb(II) complex is not formed in this region. The large p[H] increase occurring near

Table 6.5. Initial Equilibrium Data for a 2:1 Molar Ratio of 2 Iminodiacetic Acid:1 Lead(II) at 25.0°C and μ = 0.100M

mL KOH	p[H]	mL KOH	p[H]	mL KOH	p[H]
.000	2.667	1.800	3.463	3.600	8.615
.100	2.692	1.900	3.556	3.700	8.754
.200	2.718	2.000	3.656	3.800	8.871
.300	2.745	2.100	3.766	3.900	8.976
.400	2.774	2.200	3.884	4.000	9.972
.500	2.803	2.300	4.008	4.100	9.167
.600	2.836	2.400	4.143	4.200	9.254
.700	2.871	2.500	4.284	4.300	9.338
.800	2.907	2.600	4.427	4.400	9.422
.900	2.945	2.700	4.588	4.500	9.506
1.000	3.985	2.800	4.768	4.600	9.586
1.100	3.029	2.900	4.982	4.700	9.668
1.200	3.077	3.000	5.246	4.800	9.746
1.300	3.126	3.100	5.663	4.900	9.827
1.400	3.182	3.200	6.684	5.000	9.906
1.500	3.243	3.300	7.856	5.100	9.982
1.600	3.308	3.400	8.223	5.200	10.063
1.700	3.382	3.500	8.448	5.300	10.141

a = 1.5 suggests completion of 1:1 complex formation. The buffer region subsequent to a = 1.5 continues beyond a = 2.0 and is indicative of dissociation of coordinated water to form coordinated OH^-, while leaving the question of PbL_2 formation open till the calculation is completed. The initial value taken for log K_{ML} was 7.0, for log K_{ML_2} 3.0 and log K^{-H}_{MLOH} -9.0. The rationale used was the observation that for other metal ions log K_{ML} values range in the vicinity of ca. 5.5-10.5, while the second stepwise constants are much smaller. The initial hydrolysis constant was chosen as approximately -9, on the basis of the p[H] at $\underline{a}$ = 2.75. Although these estimates turn out to be reasonably close, the program would have worked successfully and the calculated *vs.* estimated values would have converged, even if the estimates had been off by as much as 5 log units. The computation of stability constants from the data in Table 6.5 is given in Table 6.6.

Table 6.6. Results of the Pb(II)-IDA Equilibrium Computations

Program BEST vers. 05/17/88

INITIAL VOLUME	50.00000
NORMALITY OF BASE	.10240
MILLIMOLES ACID	-.03301
NUMBER DATA POINTS	54
PH CORR. INCLUDED	.000

COMPONENTS:

1 Ida	.24690	MILLIMOLES
2 Pb	.11520	MILLIMOLES
3 H	.49380	MILLIMOLES

SPECIES:

	LOG BETA			
1	.0000	1 Ida	0 Pb	0 H
2	9.3431	1 Ida	0 Pb	1 H
3	11.8922	1 Ida	0 Pb	2 H
4	.0000	0 Ida	1 Pb	0 H
5	7.3619	1 Ida	1 Pb	0 H
6	9.7812	2 Ida	1 Pb	0 H
7	-2.2145	1 Ida	1 Pb	-1 H
8	.0000	0 Ida	0 Pb	1 H
9	-13.7800	0 Ida	0 Pb	-1 H

SIGMA PH FIT= .008562

DIFFERENCE TABLE:("DIFF"=LOGBETA(I)-LOGBETA(I-1))

I	DIFF	ERROR
1	.0000	.0000
2	9.3431	.0000
3	2.5491	.0000
4	-11.8922	.0000
5	7.3619	.0632
6	2.4193	.1631
7	-11.9957	.0553
8	2.2145	.0000
9	-13.7800	.0000

	VB	A	PH	PHCALC	DIFF
1	.000	.134	2.667	2.663	.004
2	.100	.175	2.692	2.690	.002
3	.200	.217	2.718	2.718	.000
4	.300	.258	2.745	2.747	-.002
5	.400	.300	2.774	2.777	-.003
6	.500	.341	2.804	2.809	-.005
7	.600	.383	2.836	2.843	-.007
8	.700	.424	2.871	2.878	-.007
9	.800	.465	2.907	2.915	-.008
10	.900	.507	2.945	2.954	-.009
11	1.000	.548	2.985	2.996	-.011
12	1.100	.590	3.029	3.040	-.011
13	1.200	.631	3.077	3.087	-.010

Table 6.6. continued

14	1.300	.673	3.126	3.138	-.012
15	1.400	.714	3.182	3.194	-.012
16	1.500	.756	3.243	3.254	-.011
17	1.600	.797	3.308	3.320	-.012
18	1.700	.839	3.382	3.392	-.010
19	1.800	.880	3.463	3.473	-.010
20	1.900	.922	3.556	3.561	-.005
21	2.000	.963	3.656	3.659	-.003
22	2.100	1.005	3.766	3.767	-.001
23	2.200	1.046	3.884	3.883	.001
24	2.300	1.088	4.008	4.007	.001
25	2.400	1.129	4.143	4.138	.005
26	2.500	1.171	4.284	4.275	.009
27	2.600	1.212	4.427	4.418	.009
28	2.700	1.254	4.588	4.572	.016
29	2.800	1.295	4.768	4.740	.028
30	2.900	1.336	4.982	4.932	.050
31	3.000	1.378	5.246	5.169	.077
32	3.100	1.419	5.663	5.512	.151
33	3.200	1.461	6.684	6.409	.275
34	3.300	1.502	7.856	7.877	-.021
35	3.400	1.544	8.223	8.240	-.017
36	3.500	1.585	8.448	8.458	-.010
37	3.600	1.627	8.615	8.620	-.005
38	3.700	1.668	8.754	8.754	.000
39	3.800	1.710	8.871	8.870	.001
40	3.900	1.751	8.976	8.974	.002
41	4.000	1.793	9.072	9.070	.002
42	4.100	1.834	9.167	9.161	.006
43	4.200	1.876	9.254	9.248	.006
44	4.300	1.917	9.338	9.332	.006
45	4.400	1.959	9.422	9.415	.007
46	4.500	2.000	9.506	9.496	.010
47	4.600	2.042	9.586	9.577	.009
48	4.700	2.083	9.668	9.658	.010
49	4.800	2.124	9.746	9.740	.006
50	4.900	2.166	9.827	9.823	.004
51	5.000	2.207	9.906	9.909	-.003
52	5.100	2.249	9.982	9.996	-.014
53	5.200	2.290	10.063	10.085	-.022
54	5.300	2.332	10.141	10.177	-.036

The three log stability constants determined for this system correspond to the following equilibria:

$$Pb^{2+} + L^{2-} \rightleftharpoons PbL \qquad \text{Log } Q = 7.36$$

$$PbL + L^{2-} \rightleftharpoons PbL_2^{2-} \qquad \text{Log } Q = 2.42$$

$$PbL \rightleftharpoons PbLOH^- + H^+ \qquad \text{Log } Q = -9.57$$

Because the second stepwise constant is of such small magnitude, the species ML_2 is seen as a minor species. The evidence for its existence is strengthened by doing a computation assuming that it doesn't exist while refining the remaining constants. Such a computation is characterized by a poorer fit overall. In particular, the region between $\underline{a}$ = 1.5 to $\underline{a}$ = 2 is noticeably in need of improvement. As frequently occurs when considering minor species, the values of log Q's for the other two constants were changed very little by leaving ML_2 out of the computation.

This example is concluded with a pictorial representation of the interplay of all of the identified species in the system 2 IDA: 1 Pb as a function of p[H]. Figure 6.3 is the species diagram for the conditions employed in the titration and serves to point out all of the above considerations, including the conclusion that PbL_2^{2-} is a minor species.

6.4 C-BISTREN

C-BISTREN (CBT)[3] is a convenient abbreviation for an analog of the well-known, recently-reported O-BISTREN[4] cryptand. The C and O prefixes refer to the central atoms of the three 5-atom bridges between the TREN moieties of the cryptands. The coordinating properties of these ligands have been described in detail elsewhere.[3,4] Each ligand possesses eight amino groups: two acting as chain-branching atoms while the remaining are incorporated in the symmetrical bridges with the entire arrangement forming an ellipsoidal cavity which can complex up to two metal ions within the cavity. Published work on the binuclear Cu(II)-O-BISTREN complex includes the discovery that a uniquely stable μ-hydroxo bridge[3] is formed between the metal ions, and that the various cationic forms of both the protonated ligand alone and its protonated and non-protonated metal cryptate complexes trap small bridging

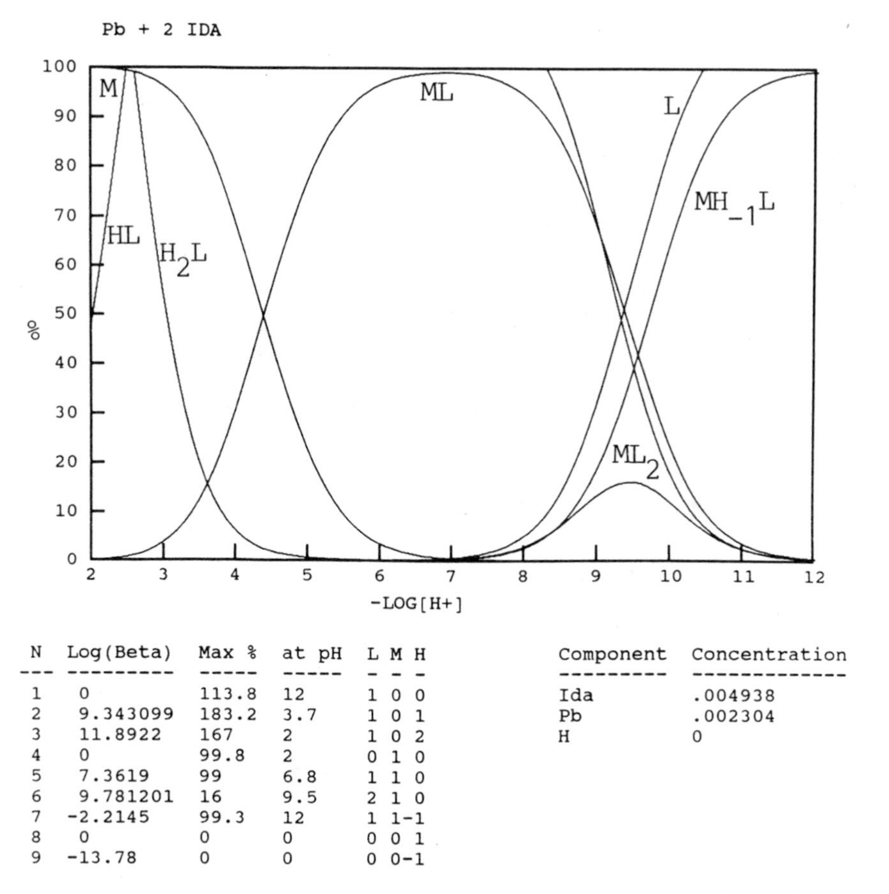

N	Log(Beta)	Max %	at pH	L	M	H
1	0	113.8	12	1	0	0
2	9.343099	183.2	3.7	1	0	1
3	11.8922	167	2	1	0	2
4	0	99.8	2	0	1	0
5	7.3619	99	6.8	1	1	0
6	9.781201	16	9.5	2	1	0
7	-2.2145	99.3	12	1	1	-1
8	0	0	0	0	0	1
9	-13.78	0	0	0	0	-1

Component	Concentration
Ida	.004938
Pb	.002304
H	0

Figure 6.3. Species Concentration Relative to T_{Pb} in a Solution having a 2:1 Molar Ratio of IDA/Pb(II), as a Function of -log $[H^+]$, μ = 0.100 M, t = 25.00°C. Species Considered are the same as in Table 6.6.

anions[5] such as Cl^-. It had been suggested[3,4,5,6] that hydrogen bonding of the bridging coordinated hydroxo group with an ether oxygen atom strongly contributes to its stability. This interpretation recently has been verified by determination of the crystal structure[7] of the binuclear cryptate $Cu_2(OH)OBTBr_3 \cdot 6H_2O$. Since it has already been shown[4] (for O-BISTREN) that perchlorate does not fit readily into the cavity, sodium perchlorate would seem to be the logical choice for the supporting electrolyte.

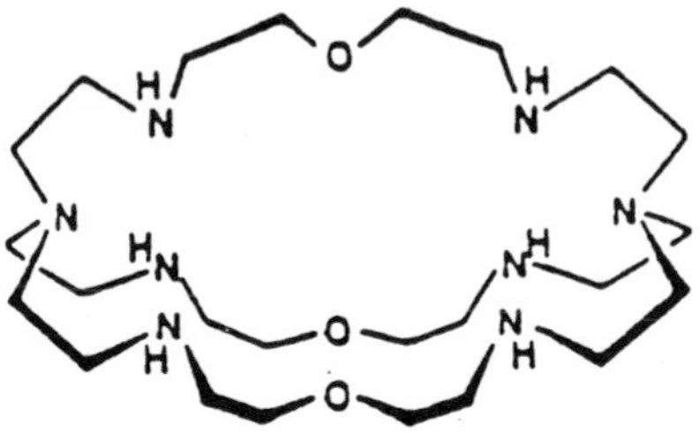

O-BISTREN (OBT)

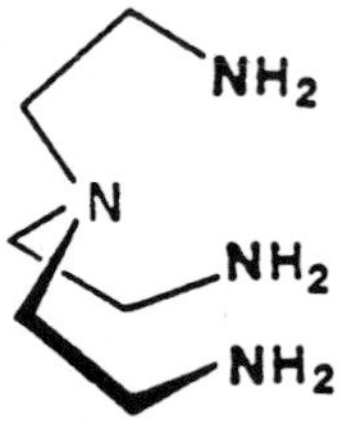

TREN

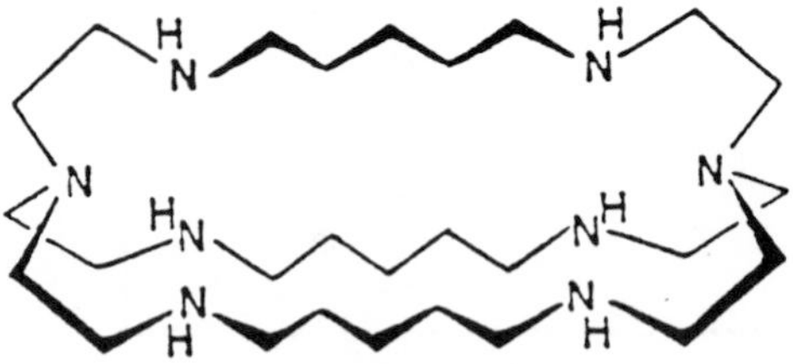

C-BISTREN (CBT)

The purpose of presenting this example is to describe in detail the stepwise procedure for determining stability constants and equilibrium constants for a complicated ligand and to go through the details leading to the quantitative description of binucleating equilibria of Cu(II) with CBT. The synthesis of this ligand is somewhat difficult and therefore it is desirable to be very economical with the material used, especially in this case. However, unlike the IDA example, a minimum of three potentiometric equilibrium data sets are required:

CBT alone

CBT with 1 copper(II) ion/ligand

CBT with 2 copper(II) ions/ligand.

It will be shown here how the required work can be accomplished with a single small sample of the ligand.

6.5 Procedure for C-BISTREN

There are three special background considerations that must be taken into account when dealing with $NaClO_4$ as supporting electrolyte.

All work must be carried out with NaOH solution as the standard base.

KCl-filled reference electrodes tend to clog in this medium.

The slight acid misbalance usually present in concentrated solutions of this salt must be determined and factored into all computations.

Carbon dioxide-free NaOH was prepared by taking a 5.50 mL sample of clear 50% NaOH solution (i.e., the clear supernatant in the presence of excess pure NaOH pellets) and diluting to 1.000 L, followed by standardization with primary-standard potassium acid phthalate. A Gran's plot indicated the absence (<0.5%) of CO_2 in the base.

The KCl solution within the calomel reference electrode was replaced with saturated NaCl. Essential drift-free behavior was established for this electrode in 24 hrs but the electrode was allowed to equilibrate for a total of 72 hrs before use. The electrodes were calibrated to read p[H] (-log $[H^+]$) with standard acid in 0.100 M $NaClO_4$ medium as described in the previous section.

CBT Alone. CBT was isolated from a concentrated HCl solution nominally as the octahydrochloride salt: $C_{27}H_{60}N_8 \cdot 8HCl$, with a formula weight of 788 Daltons. Since 0.05 millimole/50.0 mL solution is the minimum amount which can be titrated while still maintaining accuracy (corresponding to about 0.50 mL standard base/a-value), a solid sample of 41.2 mg (0.0523 mmol = 41.2 mg/788 $mg \cdot mmol^{-1}$) was placed in the vessel containing 5.00 mL of 1.00 M $NaClO_4$ and 45.00 mL doubly distilled water. The vessel was sealed and seventy two 0.050 mL increments of 0.1103 M NaOH were slowly added from a piston burette while equilibrium p[H] values were recorded with the meter. The data obtained are given in Table 6.7.

Table 6.7. Initial Titration Data for CBT-8HQ

mL	p[H]	mL	p[H]	mL	p[H]
.000	2.822	1.200	7.874	2.350	9.066
.050	2.852	1.250	7.926	2.450	9.143
.100	2.887	1.300	7.977	2.500	9.305
.150	2.924	1.350	8.027	2.550	9.387
.200	2.966	1.400	8.075	2.600	9.472
.250	3.011	1.450	8.120	2.650	9.553
.300	3.062	1.500	8.164	2.700	9.632
.350	3.122	1.550	8.207	2.750	9.707
.400	3.188	1.600	8.251	2.800	9.780
.450	3.267	1.650	8.295	2.850	9.849
.500	3.362	1.700	8.340	2.900	9.918
.550	3.483	1.750	8.385	2.950	9.976
.600	3.651	1.800	8.432	3.000	10.036
.650	3.932	1.850	8.478	3.050	10.097
.700	4.737	1.900	8.526	3.100	10.155
.750	6.580	1.950	8.575	3.150	10.210
.800	7.011	2.000	8.629	3.200	10.265
.850	7.244	2.050	8.683	3.250	10.316
.900	7.395	2.100	8.736	3.300	10.369
.950	7.515	2.150	8.794	3.350	10.421
1.000	7.607	2.200	8.855	3.400	10.472
1.050	7.685	2.250	8.921	3.450	10.521
1.110	7.756	2.300	8.991	3.500	10.569
1.150	7.817	2.350	9.066	3.550	10.613
				3.600	10.654

The first data point (p[H] = 2.822) is too high. The expected initial reading should be 2.679 and is determined by the following method. Each TREN moiety possesses three basic secondary nitrogens and a very weakly-basic bridgehead tertiary amino group. TREN itself has protonation constants of 10.17, 9.47, 8.43 and an unmeasurably low 4th value. Therefore, because of the fact that the CBT ligand possesses eight HCl's and because it has two donors of very low basicity, two nitrogens should completely release their protons in solution. Hence the computation of the theoretical initial p[H] (remembering that all of the remaining amino groups are sufficiently basic that they contribute nothing to the initial hydrogen ion concentration):

$$= -\log [0.1046 \text{ mmol}/50.00 \text{ mL}]$$

$$= 2.679$$

The observed value (p[H] 2.822) indicates the presence of less than the theoretical concentration of hydrogen ion.

$$\text{mmol}_H = \text{antilog } [-2.822] 50.00 \text{ mL}$$

$$= 0.0753 \text{ mmol}$$

This calculation indicates the presence of about 1.5 mmoles of strong acid or ca. 7.5 HCl/formula rather than the theoretical 8.0. There is a sharp rise in the p[H] profile between 0.70 and 0.75 mL base added, the exact position of which is a measure of the amount of strong acid present. A more accurate and practical way of determining the location of this potential break is to examine the p[H] differences ($diff_1$) between the successive points and the difference between these differences ($diff_2$):

mL	p[H]	$diff_1$	$diff_2$
0.65	3.932		
		0.805	
0.70	4.737		+1.038
		1.843	
0.75	6.580		-1.412
		0.431	
0.80	7.011		

Since the intervals in the volume of base are equivalent (each 0.05) the interpolation is accomplished by arithmetic proportionality of the second differences:

$$\text{Vol at steepest rise} = 0.700 \text{ mL} + 1.038/(1.038 + 1.412)(0.05)$$

$$= 0.721 \text{ mL}$$

This end-point volume corresponds to 0.0795 mmoles acidic hydrogen or a starting p[H] of 2.799, which is much closer to the observed starting p[H] of 2.822.

The above determination of the exact position of the end-point is subject to some error in this and other cases where there is a lack of symmetry (i.e., the acid side is unbuffered while the curve is shaped by an acid dissociation constant on the more basic side of the potential break). Therefore a still more accurate way to find the actual acid stoichiometry is through computer refinement of the number of strong acid groups while calculating the protonation constants. Furthermore, because the acid stoichiometry has been shown to be deficient for CBT, its exact molecular weight must also be determined. The possibility of error increases considerably when small amounts of a multidentate ligand are titrated because of the diminished buffer capacity of the ligand and the increased sensitivity of p[H] readings to small volumes of added base. Thus it is exceedingly important to match all stoichiometries to the titrant concentration. Normally, a sufficiently close value of the millimoles may be obtained if there were two sufficiently sharp breaks in the p[H] profile. However in this case, since only one such break is present and the concentration of the ligand is low, it was necessary to determine the molecular weight through the calculation of the number of millimoles most consistent with the shape of the ligand curve shown in Figure 6.4.

Table 6.8 compares the intermediate results with the final observation obtained in the determination of the protonation constants of the CBT molecule.

The empirical formula basis suffers mostly from the much too high hydrogen ion specification. The overspecification by 0.0251 millimoles (0.4184-0.3933) relative to ca. 0.05 mmoles of ligand is extremely serious and gives rise to a totally unsatisfactory calculation. The log protonation constants listed are even qualitatively meaningless. In fact, the fit is so bad that there is a region where the differences in $p[H]_{obs}-p[H]_{calc}$ are >3.0 units.

The end-point basis results are based on the amount of

ligand calculated from the formula molecular weight while the titratable protons were computed from the end-point. This result is much more satisfactory in both the sigma $p[H]_{fit}$ and in the quality of the protonation constants obtained. However the match is not perfect, especially in the acid region.

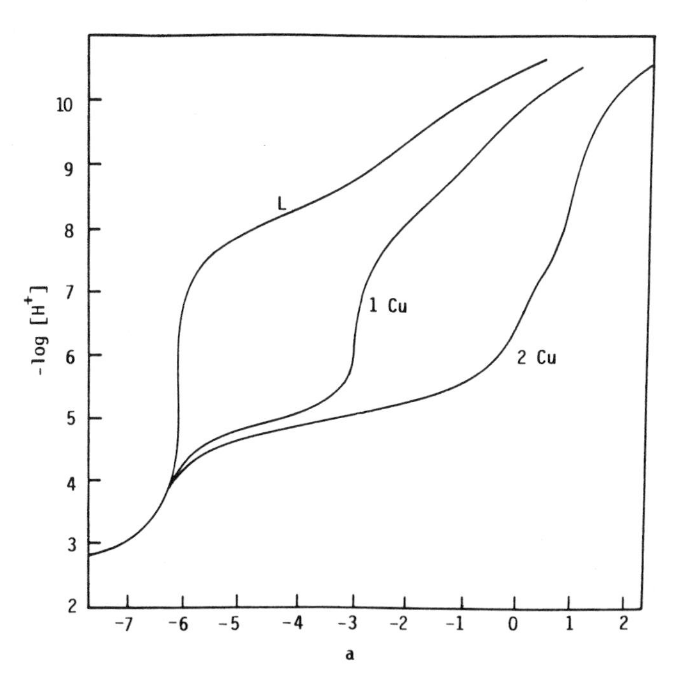

Figure 6.4. Potentiometric equilibrium of CBT curves measured in aqueous solution at 25.00°C, μ = 0.100 ($NaClO_4$) in the absence and in the presence of 1:1 and 1:2 L:M Cu(II). T_L = 1.00 x 10^{-3} M.

Upon fine tuning of both the ligand and the hydrogen ions present, the most nearly correct result was obtained and is listed in the right hand column of Table 6.8. The amount of CBT present in 41.2 mg of CBT·8HCl" is 820 Daltons (=41.2 mg/0.05023 mmol) and it possesses 7.50 equivalents of HCl (=0.37867 mmol H^+/0.05023 mmol). Since the nominal formula weight of CBT·7.5HCl is only 770 Daltons the difference (50 Daltons) is ascribable within experimental error to the presence of three waters of crystallization.

Table 6.8. Results of the Refinement of the p[H] Data for 4.21 mg CBT.8HCl Leading to the Correct Values of the Molecular Weight, Ligand-Proton Stoichiometry, and Log Protonation Constants

	Empirical Formula Basis	End-Point Basis	Fine-Tuned Refinement Basis
CBT (mmole)	0.0523	0.0523	0.05023
Proton (mmole)	0.4184	0.3933	0.37867
Log K_1^H	6.71	10.63	10.35
Log K_2^H	14.30	10.01	9.88
Log K_3^H	9.17	9.07	8.87
Log K_4^H	9.03	8.38	8.38
Log K_5^H	5.39	8.17	8.14
Log K_6^H	10.94	7.79	7.72
Log Σ_{fit}	0.249	0.014	0.0057

A final look at the results is required to satisfy the "chemically-reasonable" criterion for the values listed. Consider the structure of TREN, one of the basic building blocks of CBT, and its log protonation constants: 10.17, 9.47, and 8.43 and compare these pairwise with those of CBT: (10.35, 9.88), (8.87, 8.38), and (8.14, 7.72). In the beginning stages of protonation, separation of charge is possible. Thus the 10.17 value for TREN is similar in magnitude to the average value of the numbers within the first parentheses. As charge repulsions and other factors become more important, the second pair of CBT protonation constants is lower than the second TREN constant, while the third pair is lower than the third protonation constant of TREN. One of the apparent factors involves the similarity of the range of each set of protonation constants. Since the three charges in the triprotonated

TREN molecule are known to be separated by 5 atoms it is logical to conclude that the CBT molecule may possess a similar pattern of amino group protonation, but is more strongly influenced by steric constraints on the configuration and separation of the basic amino groups in the cryptand.

Such detailed analysis reveals the power of the potentiometric method when carried out to its fullest capability. This discussion completes the ligand characterization and protonation constant determination. Now the potentiometric determination of the CBT/Cu(II) system will be described and analyzed to show the detailed reasoning in calculating the most accurate set of stability constants possible for defining quantitatively the solution behavior of this complex as a function of p[H].

Cu(II)/CBT System. Because the supply of ligand was limited, the solution from the previous experiment was reused in the following way. To the alkaline solution obtained at the end of the first determination, 3.00 mL of 0.1170 M $HClO_4$ and 2.10 mL of 0.02386 M $Cu(ClO_4)_2$ solution were added successively. The base burette was reset to zero, the cell was resealed and 2 hrs were allowed for purging with purified N_2 for removal of CO_2, with rapid stirring, from the acidic solution. Then seventy one 0.05 mL increments of 0.1103 M NaOH were slowly added from a piston burette while p[H] values were recorded. The experimental solution contained the following quantities:

0.05023 mmole of CBT
0.05011 mmole of Cu^{2+} (2.10 mL x 0.02386 M $Cu(ClO_4)_2$)
0.3315 mmole of H^+ (+0.3786 mmole initially
-0.3971 mmole NaOH (3.60 mL x 0.1103 M)
+0.3510 mmole $HClO_4$ (3.00 mL x 0.1170 M))

58.70 mL initial volume (50.00 mL + 3.60 mL NaOH + 3.00 mL $HClO_4$ + 2.1 mL $Cu(ClO_4)_2$)

The ionic strength of this solution was still approximately 0.10 M. This is checked summing the millimoles of the predominant 1:1 electrolytes contained in the initial solution:

5.000 mmole $NaClO_4$ (present initially in ligand titration)

0.397 mmole NaOH (added titrant) + 0.351 mmole $HClO_4$ (added acid)

5.397 mmole total

Thus the ionic strength is at least 0.092 M (5.397 mmol ions/58.70 mL) from the contributions of the univalent electrolytes present. After $[Cu(ClO_4)_2]$ is factored in, the calculated ionic strength becomes 0.095 (= (5.397 + $1/2(0.05011(2)^2 + 1/2(0.1002)(1)^2$). The presence of various multiply-charged ions make a further correction in the right direction well within 5% of 0.100 M. Normally, more supporting electrolyte would be added when the addition of dilute reagents causes a significant decrease in ionic strength, but it was not necessary in this case.

The 2:1 Cu(II):CBT solution was prepared from the above 1:1 solution at the end of the determination by the addition of 3.10 mL of 0.1170 M $HClO_4$ and 2.01 mL of 0.02386 M $Cu(ClO_4)_2$ solution. The base burette was again reset to zero, the cell was resealed, and 2 more hrs were allowed for purging CO_2 from the acid solution. Seventy five 0.050 mL increments of 0.1103 M NaOH were slowly added while p[H] values were recorded. This 2:1 M:L solution contains the following components:

0.05023 mmole of CBT (same as the 1:1 solution)

0.0981 mmole of Cu^{2+}

((2.10 + 2.01) mL x 0.02386 M $Cu(ClO_4)_2$)

0.3105 mmole of H^+

(+0.3338 mmole (found in the 1:1)

-0.3860 mmole NaOH (3.50 mL x 0.1103 M)

+ 0.3627 mmole $HClO_4$ (3.10 mL x 0.1170 M))

67.31 mL initial volume

(58.70 mL + 3.50 mL NaOH + 3.10 mL ($HClO_4$

+2.01 mL $Cu(ClO_4)_2$)

The potentiometric equilibrium data for these two experiments are listed in Table 6.9 and are plotted in Figure 6.4.

There are a number of useful and essential points which should be considered for any p[H] titration curve before formal computations are initiated.

An examination of the 1:1 M:L curve in Figure 6.4 shows three buffer regions. The low p[H] region is coincident with the ligand curve and therefore implies no complex formation with the metal ion. The shallow sloping region flanked by inflections at -6 and -3 moles base/mole ligand represents the formation of MH_3L^{5+} with the release of three equivalents of hydrogen ion.

$$M^{2+} + H_6L^{6+} \rightleftharpoons MH_3L^{5+} + 3\ H^+$$

The formation of the MH_3L^{5+} complex is complete at p[H] 6 and the following sloping buffer region is consistent with the subsequent stepwise neutralization of the remaining hydrogen ions on the coordinated ligand:

$$MH_3L^{5+} \rightleftharpoons MH_2L^{4+} + H^+$$

$$MH_2L^{4+} \rightleftharpoons MHL^{3+} + H^+$$

$$MHL^{3+} \rightleftharpoons ML^{2+} + H^+$$

Table 6.9. p[H] Data for 1:1 Cu(II):CBT and 2:1 Cu(II):CBT

mL	p[H]	mL	p[H]	mL	p[H]
		1:1			
.000	3.276	1.200	5.095	2.400	8.647
.050	3.346	1.250	5.130	2.450	8.750
.100	3.437	1.300	5.167	2.500	8.857
.150	3.545	1.350	5.210	2.550	8.966
.200	3.707	1.400	5.253	2.600	9.083
.250	3.930	1.450	5.303	2.650	9.203
.300	4.222	1.500	5.360	2.700	9.324
.350	4.413	1.550	5.430	2.750	9.439
.400	4.512	1.600	5.527	2.800	9.551
.450	4.579	1.650	5.700	2.850	9.650
.500	4.633	1.700	6.288	2.900	9.749
.550	4.676	1.750	6.916	2.950	9.842
.600	4.717	1.800	7.240	3.000	9.926
.650	4.754	1.850	7.453	3.050	10.009
.700	4.790	1.900	7.618	3.100	10.084
.750	4.820	1.950	7.752	3.150	10.155
.800	4.851	2.000	7.872	3.200	10.223
.850	4.882	2.050	7.982	3.250	10.288
.900	4.913	2.100	8.082	3.330	10.348
.950	4.943	2.150	8.174	3.350	10.406
1.000	4.973	2.200	8.267	3.400	10.462
1.050	5.012	2.250	8.361	3.450	10.513
1.100	5.035	2.300	8.453	3.500	10.560
1.150	5.066	2.350	8.550		
		2:1			
.000	4.147	1.250	5.019	2.500	5.925
.050	4.287	1.300	5.042	2.550	6.021
.100	4.378	1.350	5.066	2.600	6.155
.150	4.440	1.400	5.091	2.650	6.336
.200	4.489	1.450	5.116	2.700	6.585
.250	4.532	1.500	5.141	2.750	6.822
.300	4.568	1.550	5.166	2.800	7.021
.350	4.600	1.600	5.194	2.850	7.180
.400	4.630	1.650	5.221	2.900	7.337
.450	4.656	1.700	5.249	2.950	7.497
.500	4.682	1.750	5.278	3.000	7.677
.550	4.706	1.800	5.305	3.050	7.898
.600	4.730	1.850	5.335	3.100	8.172
.650	4.754	1.900	5.365	3.150	8.536
.700	4.777	1.950	5.397	3.200	8.941
.750	4.799	2.000	5.431	3.250	9.278
.800	4.821	2.050	5.463	3.300	9.548
.850	4.843	2.100	5.501	3.350	9.771
.900	4.864	2.150	5.536	3.400	9.951
.950	4.887	2.200	5.578	3.450	10.099
1.000	4.909	2.250	5.621	3.500	10.217
1.050	4.930	2.300	5.669	3.550	10.317
1.100	4.952	2.350	5.720	3.600	10.399
1.150	4.974	2.400	5.777	3.650	10.470
1.200	4.997	2.450	5.843	3.700	10.533

The shape of the 2:1 M:L curve is quite different from that of the 1:1 M:L p[H] profile. The main buffer region is six equivalents long and is followed by an additional buffer region at higher p[H], indicating a single hydrogen ion neutralization beyond that required to neutralize the hydrogen ions of the hexaprotonated ligand. At this point it is safe to assume that the early part of the long buffer region represents, as in the 1:1 M:L case, the formation of MH_3L^{5+} and the neutralization of 3 H^+ while concomitantly the M_2L^{4+} also forms with the release and neutralization of three more equivalents of hydrogen ion.

$$M^{2+} + H_6L^{6+} \rightleftharpoons MH_3L^{5+} + 3\ H^+$$

$$M^{2+} + MH_3L^{5+} \rightleftharpoons M_2L^{4+} + 3\ H^+$$

The final buffer region near p[H] 7 is indicative of μ-hydroxo bridge formation between the metal ions within the cavity of the CBT molecule, represented by the equilibrium

$$M_2L^{4+} \rightleftharpoons M_2LOH^{3+} + H^+$$

The presence of other less obvious species was also determined during the calculation described below.

Computation of Cu-CBT Stability Constants. The computation of the stability constants was accomplished in the following way:

Values of an initial set of four overall stability constants defining the formation of the mononuclear species ML^{2+}, MHL^{3+}, MH_2L^{4+}, and MH_3L^{5+} and values of an initial set of two overall stability constants defining the formation of the dinuclear species M_2L^{4+} and M_2LOH^{3+} were estimated.

The mononuclear set of overall constants was first refined by the use of the 1:1 M:L data (Table 6.9, top) with the second set fixed.

The second, binuclear set was refined by the use of the 2:1 M:L p[H] data (Table 6.9, bottom), with the first set fixed.

The corresponding refinements were repeated until further improvement was not possible.

Since the 2:1 p[H] profile was not totally resolved in the long buffer region, the additional species M_2HL^{5+} was found to be required to bring the calculated and experimental p[H] values into line. This species was unexpected but it is reasonable. Other possible formulations that would improve σ_{fit} were not considered reasonable.

The high p[H] region at approximately p[H] 9-10 was also unsatisfactory and it was necessary to include $M_2L(OH)_2^{2+}$ as an additional hydrolyzed species.

With these two additional species, and the corresponding estimated equilibrium constants, all the equilibrium constants for the 1:1 and 2:1 p[H] profiles were refined repeatedly until further improvement did not take place.

The progress of this calculation is summarized in Table 6.10. The first line represents the starting point for the initial estimates of the overall stability constants. They were obtained in conjunction with the inspection of the p[H] profiles in Table 6.9 in the following way. With the known values for OBISTREN as a guide, the log K_{ML} was estimated as 17.0 and the successive 1:1 M:L log chelate protonation constants were estimated from the p[H] curve as 8.5, 8.0 and 7.5. The 2:1 M:L stepwise formation constant was estimated by analogy as 10.0 log units and the log dissociation constant of M_2L^{4+} to give a hydrogen ion and M_2LOH^{3+} was read from the 2:1 p[H] curve as about -7.5. Thus the initial entries of Table

6.10 were obtained by summing up the appropriate log values of the stepwise constants. Table 6.10 shows that the first 5 iterations were sufficient to get the six desired beta values.

Table 6.10. Initial Estimates, Progress of Refinement, and Final Results of the Successive Iterations using the 1:1 and 2:1 M:CBT p[H] Equilibrium Data of Table 6.9.

No.	Log β_1	Log β_2	Log β_3	Log β_4	σ_{fit}	Log β_5	Log β_6	Log β_7	Log β_8	σ_{fit}
0	17.000	25.500	33.500	41.000	.304731	27.000	------	19.500	------	.300321
1	16.415	25.697	34.356	41.925	.065379	28.895	------	21.169	------	.032001
2	15.430	25.325	34.080	41.844	.035379	28.859	------	21.146	------	.030913
3	15.453	25.327	34.078	41.847	.036847	28.860	------	21.148	------	.030753
4	15.452	25.327	34.078	41.847	.036145	28.860	------	21.148	------	.030758
5	15.451	25.327	34.078	41.847	.036146	28.860	------	21.148	------	.030757
6	15.452	25.327	34.078	41.847	.036143	28.763	33.959	21.074	10.128	.017727
7	15.727	25.475	34.142	41.836	.027647	28.750	33.987	21.057	10.114	.016867
8	15.733	25.474	34.140	41.835	.028078	28.078	33.989	21.057	10.113	.016800
9	15.732	25.474	34.139	41.835	.028110	28.749	33.990	21.057	10.113	.016792
10	15.732	25.474	34.139	41.835	.028115	28.749	33.990	21.057	10.113	.016791
11	15.228	25.414	34.120	41.783	.011103	28.743	34.153	21.149	10.325	.013891
12	15.366	25.457	34.157	41.786	.014459	28.747	34.151	21.154	10.336	.013701
13	15.379	25.461	34.160	41.784	.014527	28.749	34.150	21.157	10.340	.013647
14	15.382	25.462	34.161	41.787	.014561	28.750	34.150	21.158	10.341	.013629
15	15.382	25.462	34.161	41.787	.014567	28.811	34.119	21.262	10.505	.013950
16	15.467	25.500	34.197	41.798	.015801	28.770	34.143	21.193	10.400	.013014
17	15.419	25.476	34.174	41.791	.015022	28.758	34.149	21.174	10.369	.013401
18	15.402	25.469	34.167	41.789	.014817	28.755	34.151	21.169	10.361	.013521
19	15.402	25.467	34.166	41.788	.014762	28.754	34.151	21.167	10.357	.013551
20	15.395	25.466	34.165	41.788	.014729	28.753	34.151	21.166	10.356	.014356
21	15.390	25.465	34.164	41.788	.014715	28.753	34.151	21.165	10.354	.013579
22	15.389	25.465	34.164	41.788	.014702	28.752	34.151	21.164	10.343	.013584
23	15.389	25.465	34.163	41.787	.015686	28.752	34.151	21.164	10.351	.013587
24	15.387	25.465	34.164	41.788	.014681	28.752	34.151	21.164	10.352	.013592

Log β_1-Log β_8 are respectively associated with formation of the species ML^{2+}, MHL^{3+}, MH_2L^{4+}, MH_3L^{5+}, M_2L^{4+}, M_2HL^{5+}, M_2LOH^{3+}, and $M_2L(OH)_2^{2+}$.

The σ_{fit} was not low enough overall so that log β_6 and log β_8 were considered necessary as additional species as discussed above. A considerable improvement in σ_{fit} was realized in iterations 6-10. The final iterations 11-24 represent fine tuning of the stoichiometry of the acid present, since dealing with multiple successive additions of standard base and acid is likely to introduce some mismatch in the millimoles of H^+ calculated compared to the actual amount present. The fine-tuned adjustments were quite small, from 0.3315 to 0.3337 for the 1:1 titration data and from 0.3105 to 0.3084 for 2:1 titration data, a change of only about 0.6% in each case.

The final results of the calculations are shown in Table 6.11 for the 1:1 system and in Table 6.12 for the 2:1 system.

Table 6.11. Computed Results of the 1:1 Cu(II):CBT System Showing Stability Constants in Common with Those of Table 6.12 for the 2:1 System

```
Program BEST vers. 05/17/88
        CBT 1Cu
        INITIAL VOLUME                    58.70000
        NORMALITY OF BASE                   .11030
        MILLIMOLES ACID                     .33370
        NUMBER DATA POINTS                      71
        PH CORR. INCLUDED                     .000

 COMPONENTS:
       1 CBT          .05023  MILLIMOLES
       2 Cu(II)       .05011  MILLIMOLES
       3 proton       .00001  MILLIMOLES

 SPECIES:
   LOG BETA
 1  15.3871  1 CBT        1 Cu(II)     0 proton
 2  25.4648  1 CBT        1 Cu(II)     1 proton
 3  34.1636  1 CBT        1 Cu(II)     2 proton
 4  41.7879  1 CBT        1 Cu(II)     3 proton
 5  28.7525  1 CBT        2 Cu(II)     0 proton
 6  34.1514  1 CBT        2 Cu(II)     1 proton
 7  21.1639  1 CBT        2 Cu(II)    -1 proton
 8  10.3519  1 CBT        2 Cu(II)    -2 proton
 9    .0000  1 CBT        0 Cu(II)     0 proton
10  10.3549  1 CBT        0 Cu(II)     1 proton
11  20.2358  1 CBT        0 Cu(II)     2 proton
12  29.1050  1 CBT        0 Cu(II)     3 proton
13  37.4813  1 CBT        0 Cu(II)     4 proton
14  45.6206  1 CBT        0 Cu(II)     5 proton
15  53.3378  1 CBT        0 Cu(II)     6 proton
16    .0000  0 CBT        1 Cu(II)     0 proton
17    .0000  0 CBT        0 Cu(II)     1 proton
18 -13.7800  0 CBT        0 Cu(II)    -1 proton

        SIGMA PH FIT=   .014846

DIFFERENCE TABLE:("DIFF"=LOGBETA(I)-LOGBETA(I-1))
I      DIFF      ERROR
1    15.3871     .0000
2    10.0777     .0000
3     8.6988     .0000
4     7.6243     .0000
```

Table 6.11. continued

5	-13.0354	.0000
6	5.3989	.0000
7	-12.9875	.0000
8	-10.8120	.0000
9	-10.3519	.0000
10	10.3549	.0000
11	9.8809	.0000
12	8.8692	.0000
13	8.3763	.0000
14	8.1393	.0000
15	7.7172	.0000
16	-53.3378	.0000
17	.0000	.0000
18	-13.7800	.0000

	VB	A	PH	PHCALC	DIFF
1	.000	-6.643	3.274	3.259	.015
2	.050	-6.534	3.346	3.341	.005
3	.100	-6.424	3.437	3.441	-.004
4	.150	-6.314	3.545	3.571	-.026
5	.200	-6.204	3.707	3.756	-.049
6	.250	-6.094	3.930	4.049	-.119
7	.300	-5.985	4.222	4.333	-.111
8	.350	-5.875	4.413	4.466	-.053
9	.400	-5.765	4.512	4.545	-.033
10	.450	-5.655	4.579	4.603	-.024
11	.500	-5.545	4.633	4.650	-.017
12	.550	-5.436	4.676	4.690	-.014
13	.600	-5.326	4.717	4.726	-.009
14	.650	-5.216	4.754	4.759	-.005
15	.700	-5.106	4.790	4.790	-.000
16	.750	-4.997	4.820	4.820	-.000
17	.800	-4.887	4.851	4.849	.002
18	.850	-4.777	4.882	4.878	.004
19	.900	-4.667	4.913	4.906	.007
20	.950	-4.557	4.943	4.935	.008
21	1.000	-4.448	4.973	4.963	.010
22	1.050	-4.338	5.012	4.993	.019
23	1.100	-4.228	5.035	5.023	.012
24	1.150	-4.118	5.066	5.055	.011
25	1.200	-4.008	5.095	5.088	.007
26	1.250	-3.899	5.130	5.124	.006
27	1.300	-3.789	5.167	5.163	.004
28	1.350	-3.679	5.210	5.205	.005
29	1.400	-3.569	5.253	5.253	.000
30	1.450	-3.459	5.303	5.308	-.005
31	1.500	-3.350	5.360	5.376	-.016
32	1.550	-3.240	5.430	5.467	-.037
33	1.600	-3.130	5.527	5.612	-.085
34	1.650	-3.020	5.700	5.998	-.298
35	1.700	-2.910	6.288	6.682	-.394
36	1.750	-2.801	6.916	7.029	-.113
37	1.800	-2.691	7.240	7.254	-.014
38	1.850	-2.581	7.453	7.430	.023
39	1.900	-2.471	7.618	7.578	.040
40	1.950	-2.361	7.752	7.711	.041
41	2.000	-2.252	7.872	7.832	.040
42	2.050	-2.142	7.982	7.947	.035
43	2.100	-2.032	8.082	8.055	.027
44	2.150	-1.922	8.174	8.159	.015

Table 6.11. continued

45	2.200	-1.812	8.267	8.260	.007
46	2.250	-1.703	8.361	8.358	.003
47	2.300	-1.593	8.453	8.456	-.003
48	2.350	-1.483	8.550	8.553	-.003
49	2.400	-1.373	8.647	8.651	-.004
50	2.450	-1.263	8.750	8.751	-.001
51	2.500	-1.154	8.857	8.854	.003
52	2.550	-1.044	8.966	8.961	.005
53	2.600	-.934	9.083	9.073	.010
54	2.650	-.824	9.203	9.188	.015
55	2.700	-.715	9.324	9.305	.019
56	2.750	-.605	9.439	9.422	.017
57	2.800	-.495	9.551	9.536	.015
58	2.850	-.385	9.650	9.644	.006
59	2.900	-.275	9.749	9.745	.004
60	2.950	-.166	9.842	9.839	.003
61	3.000	-.056	9.926	9.926	.000
62	3.050	.054	10.009	10.007	.002
63	3.100	.164	10.084	10.083	.001
64	3.150	.274	10.155	10.154	.001
65	3.200	.383	10.223	10.220	.003
66	3.250	.493	10.288	10.282	.006
67	3.300	.603	10.348	10.341	.007
68	3.350	.713	10.406	10.396	.010
69	3.400	.823	10.462	10.447	.015
70	3.450	.932	10.513	10.496	.017
71	3.500	1.042	10.560	10.542	.018

Table 6.12. Computed Results of the 2:1 Cu(II):CBT System Showing Stability Constants in Common with Those of Table 6.11 for the 1:1 System

```
Program BEST vers. 05/17/88
        CBT 2Cu
        INITIAL VOLUME                   67.31000
        NORMALITY OF BASE                  .11030
        MILLIMOLES ACID                    .30844
        NUMBER DATA POINTS                     75
        PH CORR. INCLUDED                    .000

 COMPONENTS:
       1 CBT          .05023  MILLIMOLES
       2 Cu(II)       .09806  MILLIMOLES
       3 proton       .00001  MILLIMOLES

 SPECIES:
   LOG BETA
 1  15.3871  1 CBT        1 Cu(II)    0 proton
 2  25.4648  1 CBT        1 Cu(II)    1 proton
 3  34.1636  1 CBT        1 Cu(II)    2 proton
 4  41.7879  1 CBT        1 Cu(II)    3 proton
 5  28.7525  1 CBT        2 Cu(II)    0 proton
 6  34.1514  1 CBT        2 Cu(II)    1 proton
 7  21.1639  1 CBT        2 Cu(II)   -1 proton
 8  10.3519  1 CBT        2 Cu(II)   -2 proton
```

Table 6.12. continued

9	.0000	1	CBT	0	Cu(II)	0	proton
10	10.3549	1	CBT	0	Cu(II)	1	proton
11	20.2358	1	CBT	0	Cu(II)	2	proton
12	29.1050	1	CBT	0	Cu(II)	3	proton
13	37.4813	1	CBT	0	Cu(II)	4	proton
14	45.6206	1	CBT	0	Cu(II)	5	proton
15	53.3378	1	CBT	0	Cu(II)	6	proton
16	.0000	0	CBT	1	Cu(II)	0	proton
17	.0000	0	CBT	0	Cu(II)	1	proton
18	-13.7800	0	CBT	0	Cu(II)	-1	proton

SIGMA PH FIT= .013591

DIFFERENCE TABLE:("DIFF"=LOGBETA(I)-LOGBETA(I-1))

I	DIFF	ERROR
1	15.3871	.0000
2	10.0777	.0000
3	8.6988	.0000
4	7.6243	.0000
5	-13.0354	.0000
6	5.3989	.0000
7	-12.9875	.0000
8	-10.8120	.0000
9	-10.3519	.0000
10	10.3549	.0000
11	9.8809	.0000
12	8.8692	.0000
13	8.3763	.0000
14	8.1393	.0000
15	7.7172	.0000
16	-53.3378	.0000
17	.0000	.0000
18	-13.7800	.0000

	VB	A	PH	PHCALC	DIFF
1	.000	-6.141	4.147	3.952	.195
2	.050	-6.031	4.287	4.209	.078
3	.100	-5.921	4.378	4.355	.023
4	.150	-5.811	4.440	4.439	.001
5	.200	-5.701	4.489	4.498	-.009
6	.250	-5.592	4.532	4.544	-.012
7	.300	-5.482	4.568	4.582	-.014
8	.350	-5.372	4.600	4.615	-.015
9	.400	-5.262	4.630	4.645	-.015
10	.450	-5.152	4.656	4.672	-.016
11	.500	-5.043	4.682	4.697	-.015
12	.550	-4.933	4.706	4.721	-.015
13	.600	-4.823	4.730	4.744	-.014
14	.650	-4.713	4.754	4.765	-.011
15	.700	-4.603	4.777	4.787	-.010
16	.750	-4.494	4.799	4.808	-.009
17	.800	-4.384	4.821	4.828	-.007
18	.850	-4.274	4.843	4.848	-.005
19	.900	-4.164	4.864	4.869	-.005
20	.950	-4.054	4.887	4.889	-.002
21	1.000	-3.945	4.909	4.909	-.000
22	1.050	-3.835	4.930	4.930	.000
23	1.100	-3.725	4.952	4.950	.002
24	1.150	-3.615	4.974	4.971	.003
25	1.200	-3.505	4.997	4.993	.004
26	1.250	-3.396	5.019	5.015	.004
27	1.300	-3.286	5.042	5.037	.005

Table 6.12. continued

28	1.350	-3.176	5.066	5.060	.006
29	1.400	-3.066	5.091	5.083	.008
30	1.450	-2.957	5.116	5.108	.008
31	1.500	-2.847	5.141	5.133	.008
32	1.550	-2.737	5.166	5.159	.007
33	1.600	-2.627	5.194	5.185	.009
34	1.650	-2.517	5.221	5.213	.008
35	1.700	-2.408	5.249	5.242	.007
36	1.750	-2.298	5.278	5.271	.007
37	1.800	-2.188	5.305	5.302	.003
38	1.850	-2.078	5.335	5.333	.002
39	1.900	-1.968	5.365	5.366	-.001
40	1.950	-1.859	5.397	5.400	-.003
41	2.000	-1.749	5.431	5.435	-.004
42	2.050	-1.639	5.463	5.471	-.008
43	2.100	-1.529	5.501	5.509	-.008
44	2.150	-1.419	5.536	5.549	-.013
45	2.200	-1.310	5.578	5.590	-.012
46	2.250	-1.200	5.621	5.634	-.013
47	2.300	-1.090	5.669	5.682	-.013
48	2.350	-.980	5.720	5.733	-.013
49	2.400	-.870	5.777	5.788	-.011
50	2.450	-.761	5.843	5.850	-.007
51	2.500	-.651	5.925	5.919	.006
52	2.550	-.541	6.021	6.000	.021
53	2.600	-.431	6.155	6.098	.057
54	2.650	-.321	6.336	6.222	.114
55	2.700	-.212	6.585	6.394	.191
56	2.750	-.102	6.822	6.644	.178
57	2.800	.008	7.021	6.928	.093
58	2.850	.118	7.180	7.169	.011
59	2.900	.228	7.337	7.373	-.036
60	2.950	.337	7.497	7.559	-.062
61	3.000	.447	7.677	7.743	-.066
62	3.050	.557	7.898	7.941	-.043
63	3.100	.667	8.172	8.175	-.003
64	3.150	.777	8.536	8.491	.045
65	3.200	.886	8.941	8.963	-.022
66	3.250	.996	9.278	9.435	-.157
67	3.300	1.106	9.548	9.722	-.174
68	3.350	1.216	9.771	9.908	-.137
69	3.400	1.326	9.951	10.044	-.093
70	3.450	1.435	10.099	10.151	-.052
71	3.500	1.545	10.217	10.240	-.023
72	3.550	1.655	10.317	10.315	.002
73	3.600	1.765	10.399	10.380	.019
74	3.650	1.874	10.470	10.439	.031
75	3.700	1.984	10.533	10.491	.042

The results of the computations shown in the two preceding tables (6.11 and 6.12) are remarkable considering the small amount of sample employed, the small amounts of standard base added, and the amount of information obtained in the form of equilibrium constants. In Table 6.11 there are two small regions near p[H] 4 and 6 containing several ill-fitting points (i.e., diff > 0.1 p[H] unit) and in Table 6.12 there are three such regions. All correspond to the steep inflection regions visible in the p[H] profiles of the Cu(II) systems in Figures 6.4. Such behavior is normal, especially in such a dilute system, and is no cause for concern. The program BEST is set up so as to put much less weight in such steeply sloping data than to the shallow-slope buffer regions.

Table 6.13 contains a summary of the thermodynamic constants obtained, written in a more conceptual stepwise fashion, together with the corresponding values obtained with the CBT analog OBISTREN, both determined in perchlorate medium of 0.10 M ionic strength at 25°C.

The stability constants listed seem somewhat abstract, taken by themselves. However they are thermodynamic parameters, which apply to the reaction conditions, 25.0°C and $\mu = 0.100$ in aqueous solution in 0.100 M $NaClO_4$, and they relate the concentrations of the metal complex species formed by various concentrations of CBT and Cu^{2+}. Figures 6.5 and 6.6 illustrate the concentrations of metal ion containing species for 1:1 M:L and 2:1 M:L millimolar solutions, respectively.

It is important to realize why it was necessary to refine 1:1 and 2:1 constants from their respective 1:1 and 2:1 p[H] data profiles. Note that the 1:1 (Figure 6.5) solutions contain predominating concentrations of MH_3L, MH_2L, and MHL at various pH values, with only a small (ca. 10%) concentration of ML at pH 10. Also note that this Figure shows the presence of 2:1 species in quite significant concentrations, whereas the 2:1 M:L (Figure 6.6) shows the presence of mainly 2M:1L species after the concentration of mononuclear MH_3L initially formed diminishes. However, throughout the pH range 7 to 11, low concentration (<5%) of mononuclear species are indicated. The β's were calculated for those complexes which predominate in their respective solutions, thus employing the most accurate calculation procedure. The refinement of 1:1 and 2:1 constants were carried out with both sets of data because of

Table 6.13. Log Stability Constants of CBT and OBT at 25.0°C and μ =0.100 M ($NaClO_4$)

Quotient	CBT	BT
$\frac{[ML^{2+}]}{[M^{2+}][L]}$	15.39	17.36
$\frac{[MHL^{3+}]}{[ML^{2+}][H^+]}$	10.08	8.13
$\frac{[MH_2L^{4+}]}{[MHL^{3+}][H^+]}$	8.70	7.51
$\frac{[MH_3L^{5+}]}{[MH_2L^{4+}][H^+]}$	7.62	5.36
$\frac{[M_2L^{4+}]}{[ML^{2+}][M^{2+}]}$	13.36	9.96
$\frac{[M_2HL^{5+}]}{[M_2L^{4+}][M^{2+}]}$	5.40	-
$\frac{[M_2LOH^{3+}][H^+]}{[M_2L^{4+}]}$	-7.59	-2.24
$\frac{[M_2L(OH)_2^{2+}][H^+]}{[M_2LOH^{3+}]}$	-10.81	-

the presence of both types of species in each solution, thus taking into account the presence of all species during the refinement.

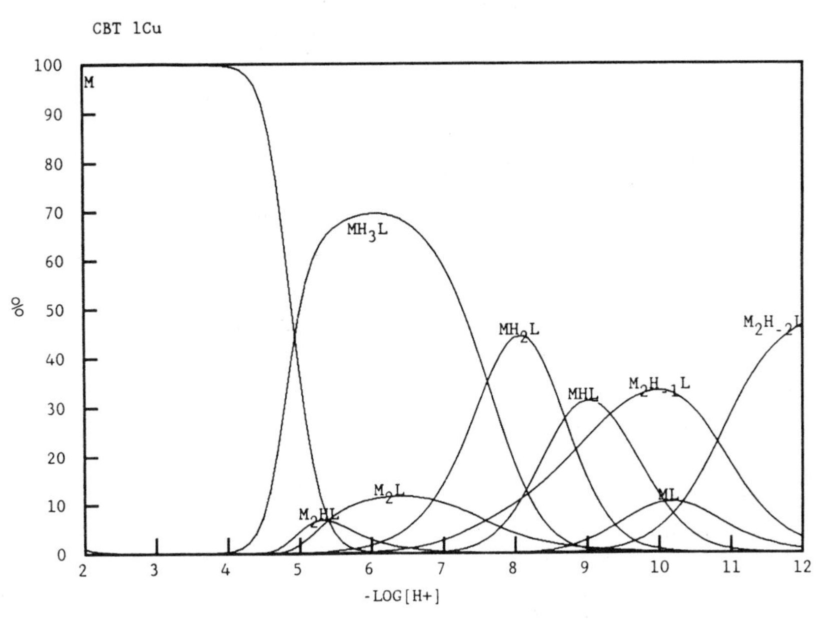

N	Log(Beta)	Max %	at pH	L M H
1	15.3871	10.4	10.2	1 1 0
2	25.4648	31.1	9	1 1 1
3	34.1636	44.4	8.1000	1 1 2
4	41.7879	69.6	6	1 1 3
5	28.7525	11.6	6.3	1 2 0
6	34.1514	6.8	5.3	1 2 1
7	21.1639	33.2	10	1 2-1
8	10.3519	46.4	12	1 2-2
9	0	48.6	12	1 0 0
10	10.3549	18.5	10.2	1 0 1
11	20.2358	18.8	9.5	1 0 2
12	29.105	9.2	8.8	1 0 3
13	37.4813	5.5	8.3999	1 0 4
14	45.6206	634	2	1 0 5
15	53.3378	3276.7	2.8	1 0 6
16	0	99.7	2	0 1 0
17	0	0	0	0 0 1
18	-13.78	0	0	0 0-1

Component	Concentration
CBT	.0010046
Cu(II)	.0010022
proton	0

Figure 6.5. Species Concentration Relative to T_{CBT} in a Solution having a 1:1 Molar Ratio of CBT/Cu(II), as a Function of -log $[H^+]$, μ = 0.100 M, t = 25.00°C.

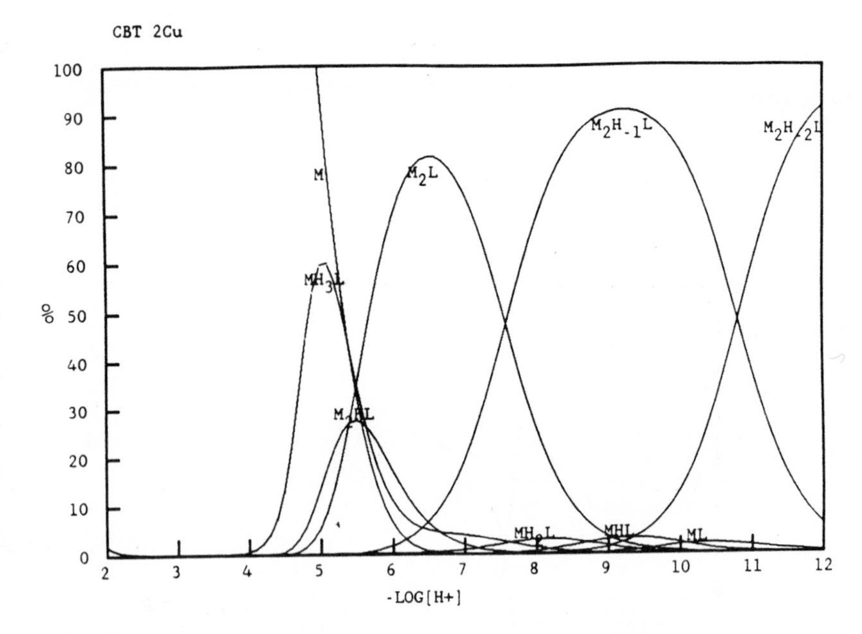

N	Log(Beta)	Max %	at pH	L	M	H
1	15.3871	2	10.3	1	1	0
2	25.4648	3.1	9.2000	1	1	1
3	34.1636	2.9	8	1	1	2
4	41.7879	59.8	5.1	1	1	3
5	28.7525	81.5	6.5	1	2	0
6	34.1514	27.6	5.5	1	2	1
7	21.1639	90.8	9.2000	1	2	-1
8	10.3519	91.5	12	1	2	-2
9	0	2.2	12	1	0	0
10	10.3549	.3	10.3	1	0	1
11	20.2358	.1	9.5	1	0	2
12	29.105	0	0	1	0	3
13	37.4813	0	0	1	0	4
14	45.6206	634	2	1	0	5
15	53.3378	3276.7	2.8	1	0	6
16	0	195.2	2.1	0	1	0
17	0	0	0	0	0	1
18	-13.78	0	0	0	0	-1

Component	Concentration
CBT	.0010046
Cu(II)	.0019612
proton	0

Figure 6.6. Species Concentration Relative to T_{CBT} in a Solution having a 1:2 Molar Ratio of CBT/Cu(II), as a Function of -log $[H^+]$, μ = 0.100 M, t = 25.00°C.

References

1. Gran, G. Analyst, **1952**, 77, 661.
2. Smith, R. M.; Martell, A. E. "Critical Stability Constants" Vol.5; Plenum: New York, 1982.
3. Motekaitis, R. J.; Martell, A. E.; Murase, I.; Lehn, J. M.; Hosseini W. M. Inorg. Chem., in press.
4. Motekaitis, R. J.; Martell, A. E.; Lehn, J. M.; Watanabe, E. I. Inorg. Chem., **1982**, 21, 4253.

5. Motekaitis, R. J.; Martell, A. E.; Dietrich, B.; Lehn, J. M. Inorg. Chem., **1984**, 23, 1588.
6. Motekaitis, R. J.; Martell, A. E.; Murase, I. Inorg. Chem., **1986**, 25, 938.
7. Motekaitis, R. J.; Rudolf, P.; Martell, A. E.; Clearfield, A. Inorg. Chem., in press.

7

MACROSCOPIC AND MICROSCOPIC CONSTANTS

7.1 Some Definitions and Concepts

The combined effects of all acid and basic functional groups in a multidentate ligand determine the hydrogen concentration in equilibrium with that ligand. Protonation constants as customarily determined and reported are called macroscopic constants because the actual donor sites where the protons reside are not specified and may not be unique. Thus a solution "species" such as HL may in fact consist of several HL species with the proton occupying a different basic site in each of the species. Microscopic constants are the equilibrium constants for equilibria involving individual species in solution. When two or more such species exist simultaneously in solution, the microscopic constants may or may not be capable of being determined distinctly. In this chapter frequent reference will be made to both acid dissociation constants as well as to protonation constants. In order to avoid confusion, all symbols used herein will be defined explicitly.

A simple example is an α-amino acid with pK_a's of approximately 3 and 10.

At pH 3:

$$\underset{\text{RCHCOOH}}{\overset{\text{NH}_2}{|}} + \underset{\text{RCHCOO}^-}{\overset{^+\text{NH}_3}{|}} + \text{H}^+ \rightleftharpoons \underset{\text{CHCOOH}}{\overset{^+\text{NH}_3}{|}}$$

0.5×10^{-5}%	50%	10^{-3}M	50%
10^{-7} parts	1 part		1 part

At pH 10:

$$\underset{\text{RCHCOO}^-}{\overset{\text{NH}_2}{|}} + \text{H}^+ \rightleftharpoons \underset{\text{RCHCOO}^-}{\overset{^+\text{NH}_3}{|}} + \underset{\text{RCHCOOH}}{\overset{\text{NH}_2}{|}}$$

50%	10^{-10}M	50%	0.5×10^{-5}%
1 part		1 part	10^{-7} parts

At pH 3, the carboxyl group is 50% dissociated, except for an infinitesimal contribution (ca. 10^{-7}) for the dissociation of the protonated amino group. At pH 10 the alkylammonium group of the amino acid is 50% dissociated, except for a minor contribution (ca. 10^{-7}) from the protonated form of the carboxylate group. These minor interferences in the macroscopic (thermodynamic) protonation constants usually may be ignored without serious error, so that the thermodynamic equilibria may be visualized in terms of the molecular processes involved. However, this rationalization breaks down when the two functional groups have comparable hydrogen ion affinities. Thus when a ligand, for example, has phenolate and aliphatic amino groups with comparable protonation constants, both functional groups would make important contributions to the hydrogen ion concentration (i.e., appreciable concentrations of the acid and basic forms of both functional groups would be present in solution simultaneously). The most accurate method of determining the overall contribution of such systems is to measure the hydrogen ion concentration, and the quantity of acid or base added to produce the measured hydrogen ion concentration. Such measurements provide the quantities necessary to calculate macroscopic constants directly and unambiguously. Methods of determining the degree of dissociation of a ligand which are based on the properties of specific functional groups of the ligand should be used cautiously to avoid errors in the interpretation of the dissociation phenomena, and consequent incorrect values of the macroscopic dissociation (protonation) constants.

For example, consider a ligand having a basic phenol group with a microscopic pK of 10.0 and a phenolic group with a microscopic pK of 9.7. The scheme below points out that because of the closeness in magnitude of the intrinsic basicities of the phenolic and amino groups, the species HL is represented on the microscopic scale as consisting of two coexisting tautomers at concentrations of comparable magnitude (in this case [A]/[B] = 1/2) (= $10^{-10}/10^{-9.7}$). Dissociation of each microspecies to form L^- would be characterized by similar constants of somewhat lower magnitudes because of decreased charge on the molecule, and electronic interactions through the conjugated bonds of the molecule.

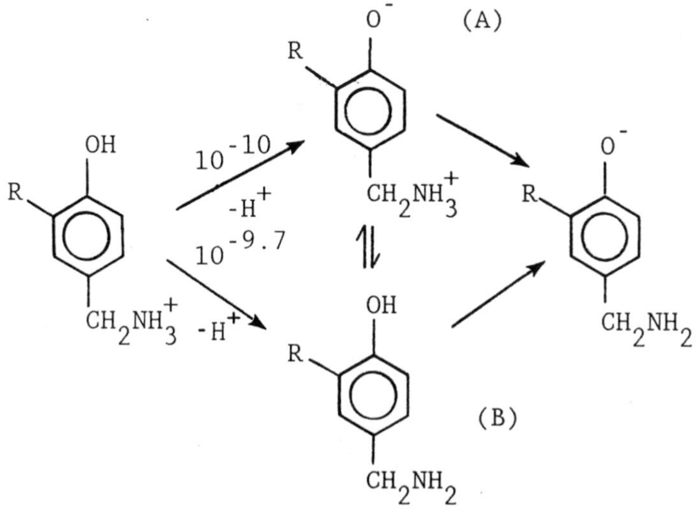

$$H_2L^+ \underset{}{\overset{K_a = 9.52}{\rightleftharpoons}} HL^o + H^+ \qquad K_D = [A]/[B] = 0.50$$

At p[H] 9.7 the concentration of B is equal to the concentration of H_2L^+. However [A]/[B] = 0.50, so that the concentration of [A] at p[H] 9.7 is $1/2[H_2L^+]$. Thus H_2L^+ is 3/5 dissociated at p[H] 9.7. Since at the true macroscopic pK $[H_2L^+]$ = [A] + [B], $[H_2L^+]$ = 3/2[B] or 3[A], and the pK is 9.7 -log 1.5 = 9.52. Thus it is seen that the macroscopic pK is lower than either of the microscopic pK's since both species contribute to the dissociation reaction and behave as a single component. Thus in general the first macroscopic acid dissociation constant K_a is the sum of the microscopic constants K_n:

$K_a = \Sigma K_n$, or in this particular example

$$K = 10^{-10} + 10^{-9.7} = 10^{-9.52}$$

On the basis of this example it is seen that:

$$K = \frac{[HL^o][H^+]}{[H_2L^+]} = \frac{([A]+[B])[H^+]}{[H_2L^+]} = \frac{[A][H^+]}{[H_2L^+]} + \frac{[B][H^+]}{[H_2L^+]}$$

Electronic spectra are often used to study the deprotonation of a phenol to form a phenolate because a phenolic group possesses an absorbance in the vicinity of 250 nm, while the conjugate phenolate group absorbs at around 300 nm, 50 nm longer wavelength. The uv probe thus is responsive only to the phenol to phenolate conversion. Applying this technique to the aminophenolate example where the first step in the dissociation reaction gives rise to two microscopic species ([A]/[B] = 0.50), it would be anticipated that only 1/3 of the phenolate uv absorbance would develop. The actual molar absorbance coefficient could be obtained for only the H_2L^+ species (and the final L^-, but not for the individual microscopic species A and B). The extinction coefficient for HL^o, if determined, would reflect the sum of A and B. Knowledge of the individual microscopic extinction coefficients, ϵ_A and ϵ_B, is required if the individual microscopic constants are to be determined. However ϵ_A and ϵ_B can only be estimated through analogies or derivatives, but cannot uniquely be determined. For example, ϵ_A could be said to be equal to ϵ_D for the derivative shown below:

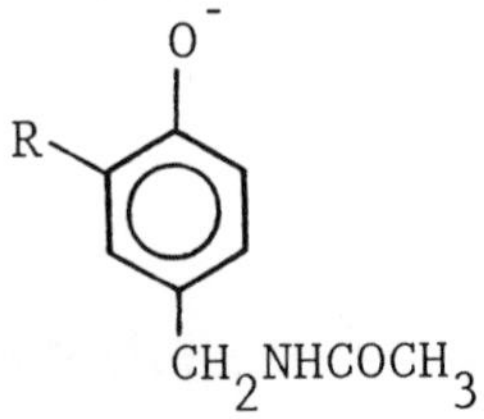

O^-

R

$CH_2NHCOCH_3$

Then, based on the ratio of observed to expected values of ϵ_{HL}, the microscopic constants can be evaluated. The calculations based on data of this type are not simple. The second step HL → L^- overlaps the first deprotonation step and therefore only an iterative value of ϵ_{HL} can be deduced through successive approximations. Similarly, the microscopic constants leading from A → L^- and B → L^- are also only accessible through estimative deduction. Molar absorbance may be used, however, in measuring the equilibrium $H_2L^+ \rightleftharpoons HL^o$, and would give the true macroscopic pK, 9.52.

The concept of microscopic constants is important and interesting, and has the advantage of being directly related to the specific functional group involved, thus enabling the investigator to interpret such parameters in terms of molecular structure. The danger lies in accepting such constants as thermodynamic constants representing the whole component, without proper justification. In some cases one cannot be certain of having taken into account all the microscopic equilibria, so that adding them up to give the macroscopic constants could be in error. In most cases this is not a problem because it is usually the individual microscopic constants that are difficult to determine, whereas the macroscopic constants can be determined directly and accurately. The following are examples of multidentate ligands that may be interpreted in terms of both macroscopic and microscopic constants.

7.2 Ionization of Tyrosine

Tyrosine, H_3L^+, **1**, is a triprotonated ligand having a single triprotonated form in strong acid, H_3L^+, and a deprotonated species in strong base, L^{2-}. However, the intermediate diprotonated and monoprotonated forms each exist as three microspecies which differ in the positions of the protons. Table 7.1 summarizes all the possibilities.

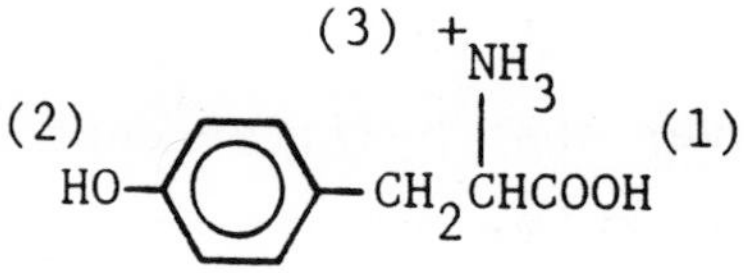

(3) $^{+}NH_3$

(2) $HO-C_6H_4-CH_2CH(NH_3^+)COOH$ (1)

1

Table 7.1. Macro Forms and Microspecies of Tyrosine

Formula	Graphic Formula	Charges		
		Carboxyl (1)	Hydroxyl (2)	Amino (3)
H_3L^+	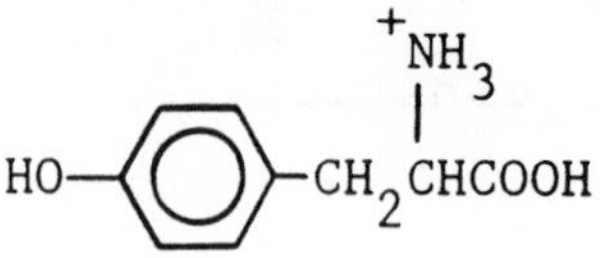 $HO-C_6H_4-CH_2CH(^+NH_3)COOH$	0	0	+
H_2L	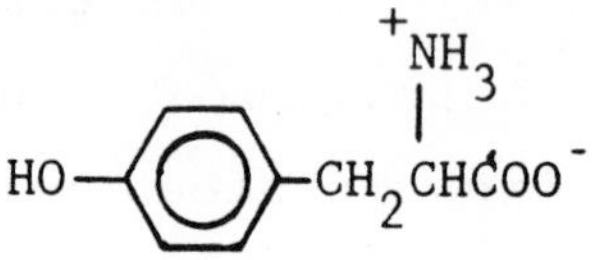 $HO-C_6H_4-CH_2CH(^+NH_3)COO^-$	-	0	+
	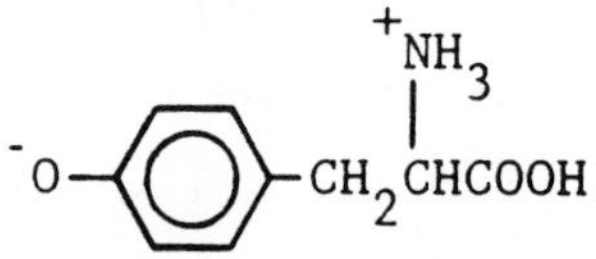 $^-O-C_6H_4-CH_2CH(^+NH_3)COOH$	0	-	+
	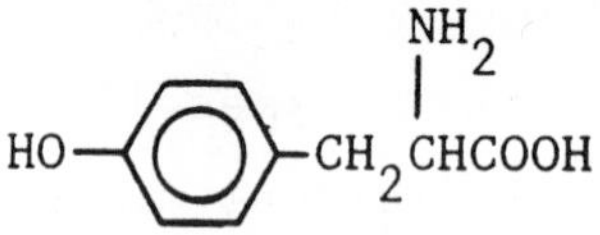 $HO-C_6H_4-CH_2CH(NH_2)COOH$	0	0	0
HL^-	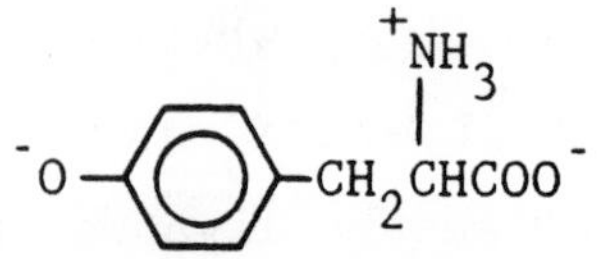 $^-O-C_6H_4-CH_2CH(^+NH_3)COO^-$	-	-	+
	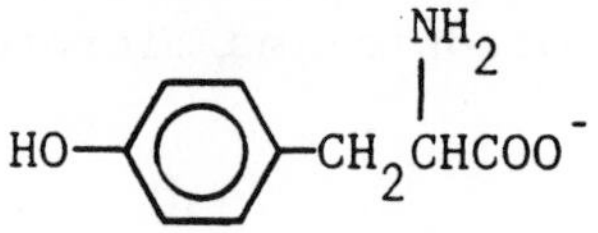 $HO-C_6H_4-CH_2CH(NH_2)COO^-$	-	0	0
	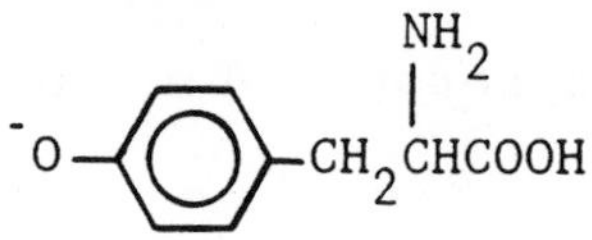 $^-O-C_6H_4-CH_2CH(NH_2)COOH$	0	-	0
L^{2-}	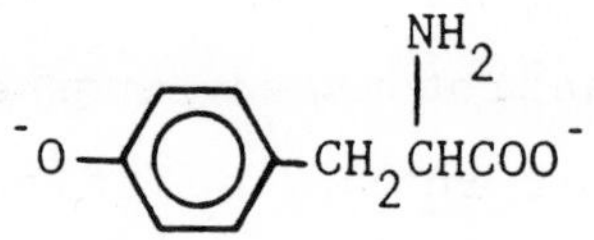 $^-O-C_6H_4-CH_2CH(NH_2)COO^-$	-	-	0

With the notation used by Martin et al.[1] for group numbers and charges, the acid-base conversion of microscopic species may be represented by Scheme 7.1

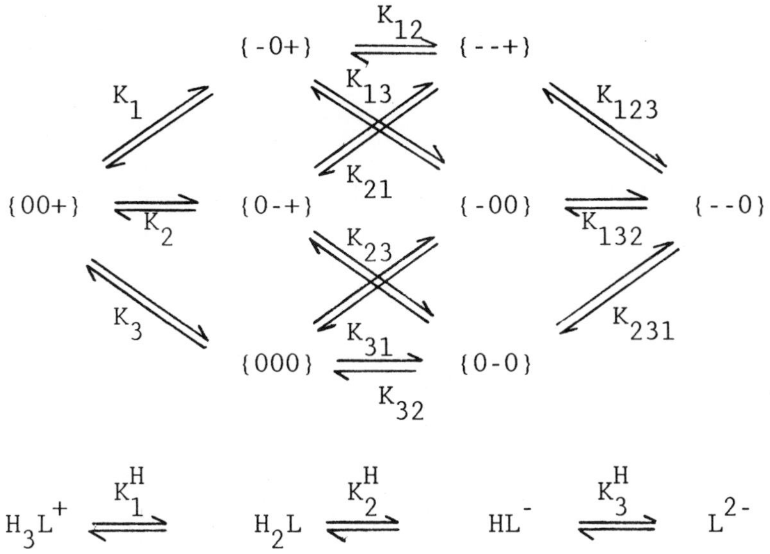

$$\{00+\} \underset{K_1,\,K_2,\,K_3}{\rightleftharpoons} \{-0+\},\ \{0-+\},\ \{000\}$$

$$\{-0+\} \underset{K_{12}}{\rightleftharpoons} \{--+\} \qquad \{-0+\} \underset{K_{13}}{\rightleftharpoons} \{-00\} \qquad \{0-+\} \underset{K_{21}}{\rightleftharpoons} \{--+\}$$

$$\{0-+\} \underset{K_{23}}{\rightleftharpoons} \{0-0\} \qquad \{000\} \underset{K_{31}}{\rightleftharpoons} \{-00\} \qquad \{000\} \underset{K_{32}}{\rightleftharpoons} \{0-0\}$$

$$\{--+\},\ \{-00\},\ \{0-0\} \underset{K_{123},\,K_{132},\,K_{231}}{\rightleftharpoons} \{--0\}$$

$$H_3L^+ \overset{K_1^H}{\rightleftharpoons} H_2L \overset{K_2^H}{\rightleftharpoons} HL^- \overset{K_3^H}{\rightleftharpoons} L^{2-}$$

COOH = 1; OH = 2; NH_3^+ = 3;

Scheme 7.1. Microscopic and Macroscopic Protonation Equilibria of Tyrosine

The macroscopic dissociation constants, determined potentiometrically by measurement of hydrogen ion concentration, are expressed in terms of the microspecies as follows:

$$K_1^H = [H^+]([-0+]+[0-+]+[000])/[00+] = K_1 + K_2 + K_3$$

$$K_2^H = [H^+[([--+]+[-00]+[0-0])/([-0+]+[0-+]+[000])$$

$$K_3^H = [H^+][--0]/([--+]+[-00]+[0-0])$$

Because the three microspecies corresponding to H_2L are removed to give way to the three corresponding to HL^-, and these are entirely removed in going to L^{2-}, the expression for macroscopic dissociation constants in terms of macroscopic constants may be simplified by using α's, the "overall" dissociation constants, whereby some of the intermediate constants cancel out ($\alpha_n^H = K_1^H \cdot K_2^H \cdot \ldots \cdot K_{n-1}^H K_n^H$)

$$\alpha_1^H = K_1 + K_2 + K_3$$

$$\alpha_2^H = K_1K_{12} + K_1K_{13} + K_2K_{23} = K_2K_{21} + K_2K_{23} + K_3K_{31}$$

$$= K_1K_{12} + K_1K_{13} + K_3K_{32} = K_1K_{12} + K_2K_{23} + K_3K_{31}$$

$$\alpha_3^H = K_1K_{12}K_{123} = K_1K_{13}K_{132} = K_2K_{21}K_{123}$$

$$= K_2K_{23}K_{231} = K_3K_{31}K_{132} = K_3K_{32}K_{231}$$

If one considers all possible microspecies, it is seen that there are six routes from H_3L^+ to HL^-. In practice, however, many of the microspecies equilibrium constants may be eliminated as being of little or no consequence. In the case of tyrosine, for example, the carboxyl group is a much stronger acid than the ammonium and phenolic groups, with the ratios K_1/K_2 and $K_1/K_3 > 10^5$. Under these conditions K_2 and K_3 may be ignored without perceptible error and $\alpha_1^H = K_1$. Thus H_2L may be considered a single species, {-0+}, and microscopic equilibria may be ignored for the $H_3L^+ \rightleftharpoons H_2L + H^+$ equilibrium (i.e., the microscopic constant K_1 is virtually equivalent to the macroscopic constant).

For the monoprotonated form, HL^-, the species {0-0} may be ignored, again because the carboxylate group is a much weaker base than the amino and phenolate group, and the per cent of protonated carboxyl in HL^- is negligible. Thus the system containing four macrospecies, two of which have three microspecies each, with a total of twelve unique microscopic constants (the K_{abc},'s in Scheme 7.1) as well as six additional derived microscopic constants ([-0+]/[0-+],

[0-+]/[000], etc.) is readily simplified to a system with four microscopic constants and only two microspecies of interest. The whole system, including the two microspecies may be defined in terms of four microscopic equilibrium constants, and three macroscopic dissociation constants (or protonation constants). Alternatively the equilibrium system may be defined by the three macroscopic constants and the distribution constant, K_d = {--+}/{-00}, which expresses the constant ratio of concentrations of microspecies, independent of p[H].

$$\{00+\} \overset{K_1}{\rightleftharpoons} \{-0+\} \rightleftharpoons \begin{matrix} \{--+\} \\ \updownarrow K_D \\ \{-00\} \end{matrix} \rightleftharpoons \{--0\}$$

$$H_3L^+ \overset{K_1}{\rightleftharpoons} H_2L \overset{K_2}{\rightleftharpoons} HL^- \overset{K_3}{\rightleftharpoons} L^{2-}$$

It is suggested here that the best way to keep microscopic and macroscopic species visualized and sorted out in the simplest and most logical fashion is to ignore the multitude of microscopic dissociation (or protonation) constants, and consider only dissociation constants for the macroscopic species with differing degrees of protonation. If the microscopic species are of interest (for visualizing molecular conformations and understanding chemical reactivities of functional groups in catalysis), they may be completely covered by the distribution constants, which express the constant ratios of their concentrations for each of the macroscopic protonated forms. Although there may be exceptions, examples of which are pointed out below, a given macroscopic species may be considered in most cases to be fully defined if the concentrations of the microspecies add up approximately to that of the macrospecies, eliminating those having insignificant concentrations (e.g., <1%).

In aqueous solution protonation equilibria are very rapid and complete equilibrium between all microspecies is generally achieved instantaneously. This is also true of measurements performed in most mixed aqueous solvents (water-methanol, water-ethanol, water-dioxane, etc.). Therefore a system of microspecies differing only in the protonation site behaves as a single compound, regardless of the number of microspecies involved. The same principle may be applied to metal complexes with alternate coordination sites. The subspecies may be considered as a unit, provided that the substitution reactions are fast relative to the measurement time.

7.3 Microscopic Protonation Equilibria of DOPA

DOPA, or 2,3-dihydroxyphenylalanine, has four protonation sites, three of which overlap extensively. A complete description of all conceivable microspecies would involve five macroscopic forms consisting of a total of fifteen microspecies (for H_3L, H_2L^- and HL^{2-}), and thirty two microscopic protonation (or dissociation) constants. A simplified treatment has been employed by Ishimitsu et al.[2] by the elimination of microscopic equilibria involving the weakly basic carboxylate group with the very strongly basic catecholate and α-amino groups ($\Delta K^H_{H_nL} > 10^6$). The resulting system, which considers microscopic equilibria between the two phenolate and the amino groups, and their protonated forms, is illustrated in Scheme 7.2. The macroscopic constants may be expressed in terms of the microscopic constants by the following relationships:

$$\alpha^H_2 = K^H_1 K^H_2 = K^H_1(K_1 + K_2 + K_3)$$

$$\alpha^H_3 = K^H_1 K^H_2 K^H_3 = K^H_1(K_1K_{12} + K_1K_{13} + K_2K_{23})$$

$$= K^H_1(K_2K_{21} + K_3K_{31} + K_3K_{32})$$

$$\alpha_4^H = K_1^H K_2^H K_3^H K_4^H = K_1^H K_1 K_{12} K_{123} = K_1^H K_2 K_{21} K_{123}$$
$$= K_1^H K_2 K_{23} K_{231} = K_1^H K_1 K_{13} K_{132}$$
$$= K_1^H K_3 K_{32} K_{231} = K_1^H K_3 K_{31} K_{132}$$

Note that with the elimination of the carboxylate group dissociation from the macroscopic equilibrium scheme, the form of Scheme 7.2 resembles closely the form of Scheme 7.1.

The determination of the microscopic equilibrium constants in these systems cannot be accomplished unambiguously, because of the interdependence of the functional groups through direct coulombic interactions and through mutual electron-withdrawing and electron-releasing effects conducted through the covalent bond system of the ligand. On the other hand Edsall, Martin and coworkers,[3,4] and others,[2] have used approximation methods based on alteration of the ligand structure, or other "reasonable" assumptions to obtain approximate values of microscopic constants. Such values, although admittedly inaccurate, provide parameters for conceptual interpretations of the nature of the molecular species that contribute to complex equilibria of multidentate ligands with hydrogen ions and metal ions.

One popular method which has been employed for the determination of microscopic constants in complex systems takes advantage of the modification of the ligand by alkylation of one or more of the acid functions. The derivative thus obtained is assumed to be sufficiently similar to the original ligand that the remaining functional groups can be measured unperturbed, assuming that the alkyl (usually methyl) group has the same electronic influence on the molecule as that of the proton which it replaces. Thus a monomethyl ether DOPA

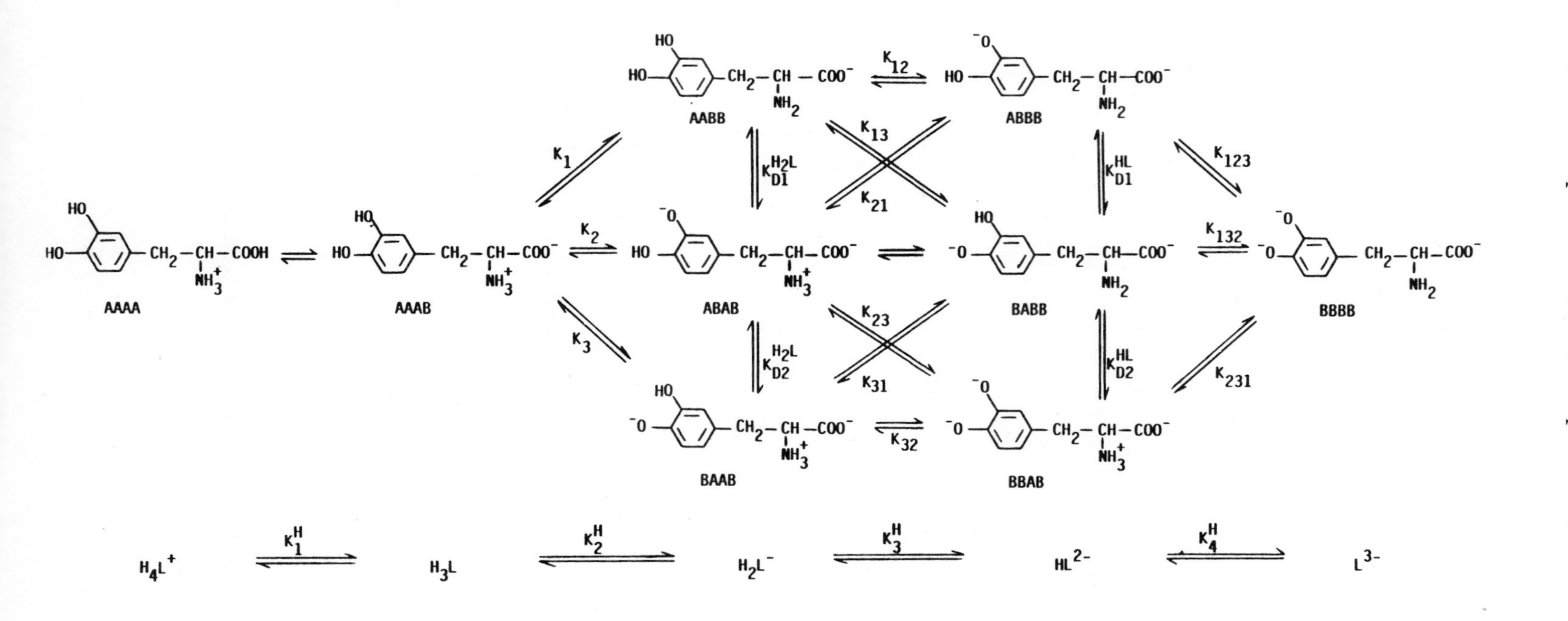

Scheme 7.2. Macroscopic and Microscopic protonation equilibria of DOPA

involving the 3- and 4-hydroxyl to give **2** or **3** would be assumed to mimic the corresponding protonated form, so that the dissociation constant of the remaining protonated phenolic group may be measured as the microscopic constant for the first dissociation (with no participation from the other phenolic group). The main difficulties with this technique, in this case and others, are (1) the assumption of the equivalence of hydrogen and alkyl groups on the dissociation properties of the remaining groups, (2) neglecting the specific intramolecular interactions between the two donor groups in both their protonated and deprotonated forms, (3) assumption that changes in charges upon derivatization are of no consequence.

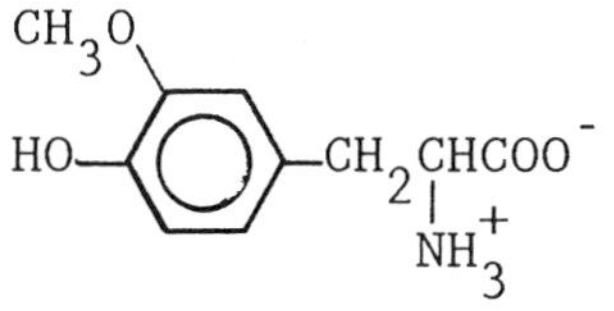

CH_3O, HO–(ring)–$CH_2CH(NH_3^+)COO^-$

2

HO, CH_3O–(ring)–$CH_2CH(NH_3^+)COO^-$

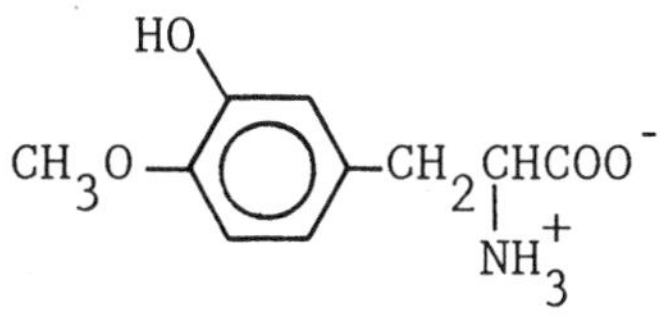

3

Another general method for the determination of microscopic constants uses a microscopic method for detecting the state (protonation or metal-coordination) of the functional group or groups being investigated. The well-known shift in the UV absorbance frequency that occurs in dissociation of a proton or a metal ion from a phenolic group may be used. Thus for tyrosine the absorption band shift is a measure of the phenolic dissociation, while determination of p[H] measures the total acid-base reaction involving hydrogen ion dissociation. For this system the use of both measurements makes possible the determination of microscopic and macroscopic dissociation constants, because of the relative noninterference by the α-amino group in the change of absorbance due to the phenolic group. Thus all important microscopic constants may be approximated unambiguously for tyrosine.

In the case of DOPA the situation is much more complex. The hydrogen ion concentration measures the overall effect of three overlapping dissociation reactions, involving two phenolic and one α-amino group. The macroscopic pK's are 2.20, 8.72, 9.78 and 13.4. Absorbance measurements indicate the phenolic group dissociations, but there is no way to discriminate between the two hydroxyl groups involved, and therefore it is not possible to determine the extent of dissociation of each group as they undergo overlapping dissociation reactions. Further, it is not possible to determine their relative contributions to the absorbance change. The usual assumption of equivalent absorbance is probably only a rough approximation. Thus we have three overlapping hydrogen ion dissociations, with two of them being phenolic dissociations, occurring in the overlapping buffer region. The microscopic dissociation constants reported for this system are based on the assumptions described above, and, therefore are recognized as rough approximations.

It should be pointed out that some degree of resolution can be achieved for the buffer region involving the two phenolic groups and one aliphatic amino group. It can be assumed that the second (whichever one that is) phenolate pK is much greater than the first one and the spectrophotometrically invariant amino group is not expected to interfere significantly with the absorbance measurements. The next two lower protonation equilibria are mixtures of amino and phenolate dissociation. Thus the absorbance changes for the two phenolate protonations can be used to qualitatively verify the involvement of phenolate in the particulate p[H] range of the neutralization equilibria.

7.4 General Comments and Conclusions

Research workers in this field should be wary of several misleading publications in which microscopic equilibria are confused with thermodynamic constants. An example is a paper on Cu(II) chelates of DOPA and related catechol amines,[5] which states that to determine stability constants, it is necessary to know the microscopic constants involving specific protonation sites of the ligand, and that the published macroscopic constants are incorrect. Such statements are themselves incorrect, and unfortunately have confused and misled many readers. Thermodynamics of complex formation determined by the measurement of hydrogen ion concentration and mass balance relationships has nothing to do with the structures of the ligands or their metal complexes, including the determination of the protonation sites of ligand and complex. The equilibrium constants express the concentration levels of the compounds defined by their chemical formulas, which are in equilibrium with each other. While each compound has a definite composition, its structure is not specified, and it may exist as several microspecies in solution having the same overall composition. The equilibrium between Cu^{2+} and DOPA in acid solution to form a diprotonated complex is a good example:

Cu^{2+} + AABB, ABAB, BAAB (from Scheme 7.2) $\rightleftharpoons$ HO(HO)C_6H_3–CH_2–CH(NH_2)(CO–O$^-$) $\cdot Cu^{2+}$

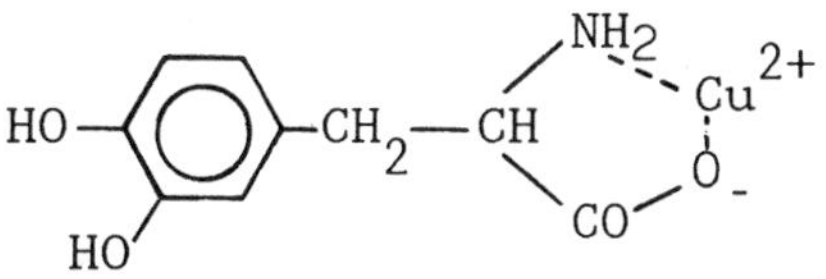

In a solution containing two molar equivalents of acid, the ligand has two functional groups in the basic form and two protonated functional groups. With the carboxylate group fully dissociated, there are two moles of hydrogen ion distributed between three basic acceptors. When Cu^{2+} combines with the amino and carboxylate groups, the phenolic groups become fully, instead of partially, protonated. It should be noted that the mixture of ligand microspecies (AABB, ABAB, BAAB) constitute a compound of unique stoichiometry and the relative concentrations of microspecies are fixed (defined by distribution constants, see Scheme 7.2). This "mixture" constitutes a single compound which does not alter in composition as the pH is changed (K_D's are pH independent), although the concentration of the compound does change with pH by virtue of its involvement in acid-base equilibria. There is nothing wrong with the redistribution of protons on the ligand on metal ion coordination, and is the natural consequence of complex formation. It is important for the investigator to understand the nature of the microspecies present, but their determination is not necessary for the measurement of protonation constants or metal binding constants.

While some investigators would like to determine the equilibrium constant for combination of metals ion with the donor groups in their basic forms, such is often not possible in real systems. For example the formation constant for the reaction of Cu^{2+} with the diprotonated form of DOPA, but with the microspecies having both phenolic groups fully protonated, would be expressed as follows:

$$Cu^{2+} + AABB^{-} \rightleftharpoons AABBCu^{+}$$

However this is simply one of three microscopic equilibria (which cannot be explicitly measured) involving formation of the diprotonated copper(II) complex. The other two microscopic reactions occur with $BAAB^{-}$ and $ABAB^{-}$. However, while it might be satisfying to visualize the magnitudes of such constants in the light of the donor groups involved, it should be realized that they are not thermodynamic constants, cannot in most cases be accurately determined, and are of theoretical interest only. It is the macroscopic constants that provide the differences in free energies between the initial and final states of complex formation.

References

1. Martin, R. B.; Edsall, J. T.; Wetlaufer, D. B.; Hollingworth, B. R. J. Biol. Chem., **1958** 233, 1429.
2. Ishimitsu, T.; Hirose, S. Talanta, **1977**, 24, 555.
3. Martin, R. B.; Edsall, J. T. J. Am. Chem. Soc., **1959**, 81, 4044.
4. Edsall, J. T.; Martin, R. B.; Hollingworth, B. R. Proc. Nat. Acad. Sci., **1958**, 44, 505.
5. Boggess, R. K.; Martin, R. B. J. Am.Chem. Soc., **1975**, 97, 3076.

DETERMINATION OF STABILITY CONSTANTS AND SPECIES CONCENTRATION OF COMPLEX SYSTEMS

8.1 A Mixed-Ligand Binuclear Dioxygen System (PXBDE + EN + Co(II) + O_2)

Potentiometry need not be limited to simple ligand, metal, ion, and proton systems. It is possible to measure the interactions of several components in much more complex systems by the methods described above. For example, recently it was of interest to study the oxygenation equilibrium of mixed ligand binuclear chelates. In particular 1,4-bis(bis(2-aminoethyl)aminomethyl)benzene hexahydrochloride, PXBDE, $(NH_2CH_2CH_2)_2NCH_2\phi CH_2N(CH_2CH_2NH_2)_2$, and EN, $NH_2CH_2CH_2NH_2$, were studied in the presence of Co^{2+} and O_2 for the purpose of computing the mixed-ligand-binuclear-Co(II)-dioxygen complex formation constant.[1,2] The following measurements were made in order to accomplish this goal:

1. PXBDE was studied potentiometrically as a function of added base to determine its six protonation constants (one p[H] profile, equations 8.1-8.6).
2. EN was studied to determine its protonation constants under the same experimental conditions (one p[H] profile, equations 8.7, 8.8).
3. PXBDE was studied with 1:1 and 2:1 molar ratios of Co(II) to ligand under nitrogen in order to determine two stepwise Co(II) formation constants and three protonation constants of the metal chelates formed. (two p[H] profiles, equations 8.9-8.13).
4. EN was studied with Co(II) under nitrogen at 1:1, 2:1 and 3:1 molar ratios of ligand to Co(II) to determine the three successive formation constants (3 p[H] profiles, equations 8.14-8.16).

5. PXBDE, EN and Co(II) were studied together at 1:2:2 molar ratio of PXBDE to EN to Co(II) under nitrogen to determine the two mixed ligand formation constants (two p[H] profiles, equations 8.17, 8.18).
6. System 5 was studied under oxygen to determine the oxygenation constant of the dinuclear mixed-ligand chelate (one p[H] profile, equation 8.19).

p[H] Profile for PXBDE

$$PXBDE + H^+ \rightleftharpoons HPXBDE^+ \qquad \beta_1 = \frac{[HPXBDE^+]}{[H^+][PXBDE]} \tag{8.1}$$

$$HPXBDE^+ + H^+ \rightleftharpoons H_2PXBDE^{2+} \qquad \beta_2 = \frac{[H_2PXBDE^{2+}]}{[H^+]^2[PXBDE]} \tag{8.2}$$

$$H_2PXBDE^{2+} + H^+ \rightleftharpoons H_3PXBDE^{3+} \qquad \beta_3 = \frac{[H_3PXBDE^{3+}]}{[H^+]^3[PXBDE]} \tag{8.3}$$

$$H_3PXBDE^{3+} + H^+ \rightleftharpoons H_4PXBDE^{4+} \qquad \beta_4 = \frac{[H_4PXBDE^{4+}]}{[H^+]^4[PXBDE]} \tag{8.4}$$

$$H_4PXBDE^{4+} + H^+ \rightleftharpoons H_5PXBDE^{5+} \qquad \beta_5 = \frac{[H_5PXBDE^{5+}]}{[H^+]^5[PXBDE]} \tag{8.5}$$

$$H_5PXBDE^{5+} + H^+ \rightleftharpoons H_6PXBDE^{6+} \qquad \beta_6 = \frac{[H_6PXBDE^{6+}]}{[H^+]^6[PXBDE]} \tag{8.6}$$

p[H] Profile for EN

$$EN + H^+ \rightleftharpoons HEN^+ \qquad \beta_1 = \frac{[HEN^+]}{[H^+][EN]} \tag{8.7}$$

$$HEN^+ + H^+ \rightleftharpoons H_2EN^{2+} \qquad \beta_2 = \frac{[H_2EN^{2+}]}{[H^+]^2[EN]} \tag{8.8}$$

2 p[H] Profiles, 1:1 $T_{PXBDE}/T_{Co^{2+}}$; 1:2 $T_{PXBDE}/T_{Co^{2+}}$

$$Co^{2+} + PXBDE \rightleftharpoons CoPXBDE^{2+} \qquad \beta_1 = \frac{[CoPXBDE^{2+}]}{[Co^{2+}][PXBDE]} \tag{8.9}$$

$$Co^{2+} + CoPXBDE^{2+} \rightleftharpoons Co_2PXBDE^{4+} \qquad \beta_2 = \frac{[Co_2PXBDE^{4+}]}{[Co^{2+}]^2[PXBDE]} \tag{8.10}$$

$$H^+ + CoPXBDE^{2+} \rightleftharpoons CoHPXBDE^{3+} \qquad \beta_3 = \frac{[CoHPXBDE^{3+}]}{[Co^{2+}][H^+][PXBDE]} \tag{8.11}$$

$$H^+ + CoHPXBDE^{3+} \rightleftharpoons CoH_2PXBDE^{4+} \qquad \beta_4 = \frac{[CoH_2PXBDE^{4+}]}{[Co^{2+}][H^+]^2[PXBDE]} \tag{8.12}$$

$$H^+ + CoH_2PXBDE^{4+} \rightleftharpoons CoH_3PXBDE^{5+} \qquad \beta_5 = \frac{[CoH_3PXBDE^{5+}]}{[Co^{2+}][H^+]^3[PXBDE]} \tag{8.13}$$

3 p[H] Profiles 1:1, 1:2, and 1:3 $[EN]/[Co^{2+}]$

$$Co^{2+} + EN \rightleftharpoons CoEN^{2+} \qquad \beta_1 = \frac{[CoEN^{2+}]}{[Co^{2+}][EN]} \tag{8.14}$$

$$EN + CoEN^{2+} \rightleftharpoons Co(EN)_2^{2+} \qquad \beta_2 = \frac{[Co(EN)_2^{2+}]}{[Co^{2+}][EN]^2} \tag{8.15}$$

$$EN + Co(EN)_2 \rightleftharpoons Co(EN)_3^{2+} \qquad \beta_3 = \frac{[Co(EN)_3^{2+}]}{[Co^{2+}][EN]^3} \tag{8.16}$$

2 p[H] Profiles, 2:1:1 and 2:1:2 $T_{Co^{2+}}/T_{PXBDE}/T_{EN}$

$$EN + Co_2PXBDE^{4+} \rightleftharpoons Co_2(PXBDE)(EN)^{4+} \tag{8.17}$$

$$\beta = \frac{[Co_2(PXBDE)(EN)^{4+}]}{[Co^{2+}]^2[PXBDE][EN]}$$

$$EN + Co_2(PXBDE)(EN)^{4+} \rightleftharpoons Co_2(PXBDE)(EN)_2^{4+} \qquad (8.18)$$

$$\beta = \frac{[Co_2(PXBDE)(EN)_2^{4+}]}{[Co^{2+}]^2[PXBDE][EN]^2}$$

p[H] Profile 2:1:2, $T_{Co^{2+}}/T_{PXBDE}/T_{EN}$ under excess O_2

$$Co_2(PXBDE)(EN)_2^{4+} + O_2 \rightleftharpoons Co_2(O_2)(PXBDE)(EN)_2^{4+} \qquad (8.19)$$

$$\beta_{O_2} = \frac{[Co_2(PXBDE)(O_2)(EN)_2^{4+}]}{[Co^{2+}]^2[O_2][PXBDE][EN]^2}$$

Although the β's for equations 8.9, 8.10 and 8.14-8.19 do not contain $[H^+]$, all of the equilibria expressed in equations 8.1-8.19 overlap, and all the p[H] profiles indicated are uniquely different, and may be used to determine the β's involved in each equation. The equilibria involved are cumulative, and the β values determined in the simpler systems are used as known quantities in solving each progressively more complex system. As the above relationships show, the p[H] profiles of ten complex systems were employed to determine nineteen equilibrium constants. The values obtained are listed in Table 8.1.

Once the β values are known, the concentrations of each of the species in these complex systems may be calculated as a function of the independent and dependent (measured) concentration variables in the system. Thus with the β values in Table 8.1, equations 8.20-8.23 may be used to calculate the concentration of each individual molecular species as a function of one of the four quantities T_M, T_H, T_L, $T_{L'}$ or $[H^+]$, where L = PXBDE, L′ = EN and M = Co, provided the remaining four are fixed. The most convenient and frequently used species plots are concentration profiles plotted as a function of p[H]. Examples of such profiles for the Co(II)-PXBDE-EN-O_2 system are given in Figure 8.1.

Table 8.1. Protonation Constants, Co(II) Complex Formation Constants, and the Oxygenation Constants for Co(II)-PXBDE-EN-Dioxygen System. (μ = 0.100 M (KNO_3); t = 25.0°C)

Quotient	Log β	Quotient	Log β
$\beta_1^H = \frac{[HPXBDE^+]}{[H^+][PXBDE]}$	10.06	$\beta_{M_2L} = \frac{[Co_2PXBDE^{4+}]}{[Co^{2+}]^2[PXBDE]}$	14.58
$\beta_2^H = \frac{[H_2PXBDE^{2+}]}{[H^+]^2[PXBDE]}$	19.77	$\beta_{MHL} = \frac{[CoHPXBDE^{3+}]}{[Co^{2+}][H^+][PXBDE]}$	18.88
$\beta_3^H = \frac{[H_3PXBDE^{3+}]}{[H^+]^3[PXBDE]}$	28.88	$\beta_{MH_2L} = \frac{[CoHPXBDE^{4+}]}{[Co^{2+}][H^+]^2[PXBDE]}$	26.60
$\beta_4^H = \frac{[H_4PXBDE^{4+}]}{[H^+]^4[PXBDE]}$	37.45	$\beta_{ML'} = \frac{[CoEN^{2+}]}{[Co^{2+}][EN]}$	5.38
$\beta_5^H = \frac{[H_5PXBDE^{5+}]}{[H^+]^5[PXBDE]}$	39.25	$\beta_{ML'_2} = \frac{[Co(EN)_2^{2+}]}{[Co^{2+}][EN]^2}$	10.40
$\beta_6^H = \frac{[H_6PXBDE^{6+}]}{[H^+]^6[PXBDE]}$	40.45	$\beta_{ML'_3} = \frac{[Co(EN)_3^{2+}]}{[Co^{2+}][EN]^3}$	13.70
$\beta_1^H = \frac{[HEN^+]}{[H^+][EN]}$	9.89	$\beta_{M_2LL'} = \frac{[Co_2(PXBDE)(EN)^{4+}]}{[Co^{2+}]^2[PXBDE][EN]}$	18.80
$\beta_2^H = \frac{[H_2EN^{2+}]}{[H^+][HEN^+]}$	16.97	$\beta_{M_2LL'_2} = \frac{[Co_2(PXBDE)(EN)_2^{4+}]}{[Co^{2+}]^2[PXBDE][EN]^2}$	23.32
$\beta_{ML} = \frac{[CoPXBDE^{2+}]}{[Co^{2+}][PXBDE]}$	NA	$\beta_{M_2LL'_2O_2} = \frac{[Co_2(PXBDE)(O_2)(EN)_2^{4+}]}{[Co^{2+}]^2[PXBDE][EN]^2P_{O_2}}$	30.03

$$T_M = \tag{8.20}$$
$$[M^{2+}] + \beta_{ML}[M^{2+}][L] + 2\beta_{M_2L}[M^{2+}]^2[L] + \beta_{MHL}[M^{2+}][H^+][L]$$
$$+ \beta_{MH_2L}[M^{2+}][H^+]^2[L] + \beta_{MH_3L}[M^{2+}][H^+]^3[L] + \beta_{ML'}[M^{2+}][L']$$
$$+ \beta_{ML'_2}[M^{2+}][L']^2 + \beta_{ML'_3}[M^{2+}][L']^3 + 2\beta_{M_2LL'}[M^{2+}]^2[L][L']$$
$$+ 2\beta_{M_2LL'_2}[M^{2+}]^2[L][L']^2 + 2\beta_{M_2L'_2O_2}[M^{2+}]^2[L][L']^2P_{O_2}$$

$$T_L = \tag{8.21}$$
$$[L] + \beta_{HL}[H^+][L] + \beta_{H_2L}[H^+]^2[L] + \beta_{H_3L}[H^+]^3[L]$$
$$+ \beta_{H_4L}[H^+]^4[L] + \beta_{H_5L}[H^+]^5[L]$$
$$+ \beta_{H_6L}[H^+]^6[L] + \beta_{ML}[M^{2+}][L]$$
$$+ \beta_{M_2L}[M^{2+}]^2[L] + \beta_{MHL}[M^{2+}][H^+][L]$$
$$+ \beta_{MH_2L}[M^{2+}][H^+]^2[L] + \beta_{MH_3L}[M^{2+}][H^+]^3[L]$$
$$+ \beta_{M_2LL'}[M^{2+}]^2[L][L'] + \beta_{M_2LL'_2}[M^{2+}]^2[L][L']^2$$
$$+ \beta_{M_2LL'_2O_2}[M^{2+}]^2[L][L']^2P_{O_2}$$

$$T_{L'} = \tag{8.22}$$
$$[L'] + \beta_{HL'}[H^+][L'] + \beta_{H_2L'}[H^+]^2[L'] + \beta_{ML'}[M^{2+}][L']$$
$$+ 2\beta_{ML'_2}[M^{2+}][L']^2 + 3\beta_{ML'_3}[M^{2+}][L']^3 + \beta_{M_2LL'}[M^{2+}]^2[L][L']$$
$$+ 2\beta_{M_2LL'_2}[M^{2+}]^2[L][L']^2 + 2\beta_{M_2LL'_2O_2}[M^{2+}]^2[L][L']^2P_{O_2}$$

$$T_{H^+} = \tag{8.23}$$
$$\beta_{HL}[H^+][L] + 2\beta_{H_2L}[H^+]^2[L] + 3\beta_{H_3L}[H^+]^3[L]$$
$$+ 4\beta_{H_4L}[H^+]^4[L] + 5\beta_{H_5L}[H^+]^5[L]$$
$$+ 6\beta_{H_6L}[H^+]^6[L] + \beta_{HL'}[H^+][L']$$
$$+ 2\beta_{H_2L'}[H^+]^2[L'] + \beta_{MHL}[M^{2+}][H^+][L]$$
$$+ 2\beta_{MH_2L}[M^{2+}][H^+]^2[L] + 3\beta_{MH_3L}[M^{2+}][H^+]^3[L]$$

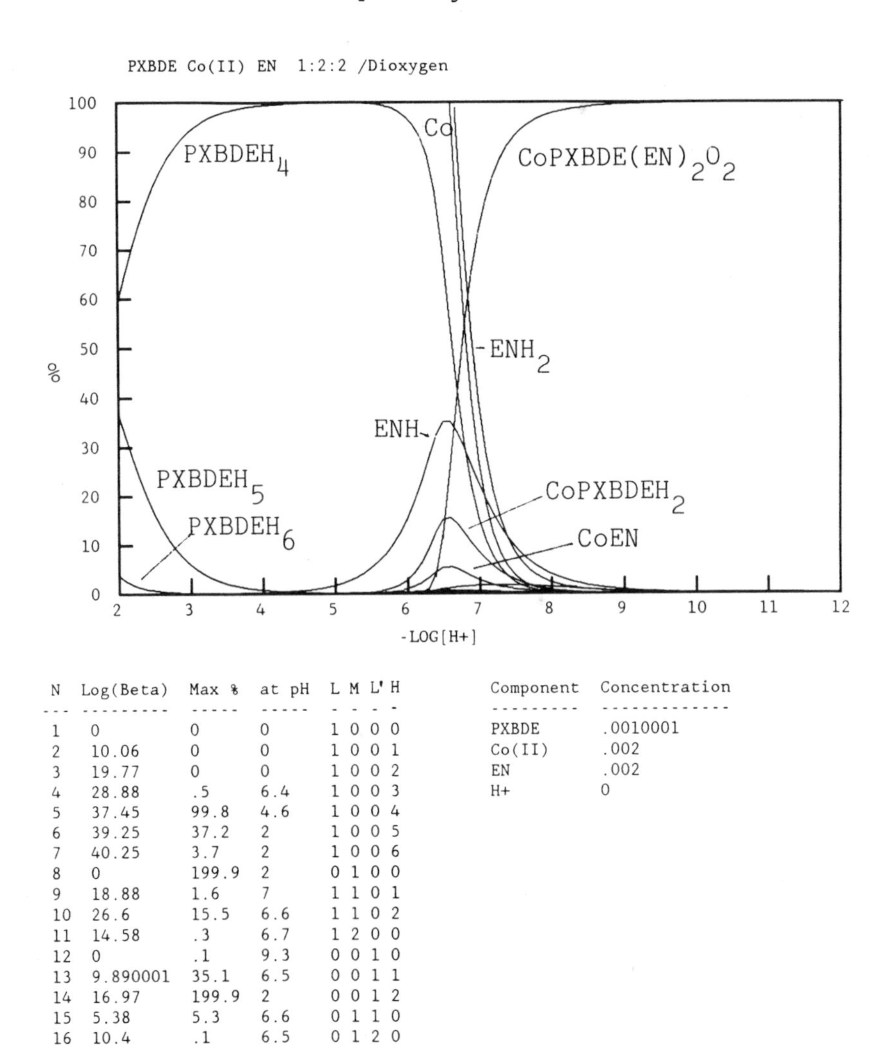

N	Log(Beta)	Max %	at pH	L M L' H
1	0	0	0	1 0 0 0
2	10.06	0	0	1 0 0 1
3	19.77	0	0	1 0 0 2
4	28.88	.5	6.4	1 0 0 3
5	37.45	99.8	4.6	1 0 0 4
6	39.25	37.2	2	1 0 0 5
7	40.25	3.7	2	1 0 0 6
8	0	199.9	2	0 1 0 0
9	18.88	1.6	7	1 1 0 1
10	26.6	15.5	6.6	1 1 0 2
11	14.58	.3	6.7	1 2 0 0
12	0	.1	9.3	0 0 1 0
13	9.890001	35.1	6.5	0 0 1 1
14	16.97	199.9	2	0 0 1 2
15	5.38	5.3	6.6	0 1 1 0
16	10.4	.1	6.5	0 1 2 0
17	13.7	0	0	0 1 3 0
18	18.8	0	0	1 2 1 0
19	23.32	0	0	1 2 2 0
20	30.03	99.9	10.3	1 2 2 0
21	0	0	0	0 0 0 1
22	-13.78	0	0	0 0 0 -1

Component	Concentration
PXBDE	.0010001
Co(II)	.002
EN	.002
H+	0

Figure 8.1. Species formed as a function of p[H] in an aqueous solution of PXBDE containing 2 moles Co(II) and EN (ethylenediamine) saturated with dioxygen. All constants used were determined at 25.0°C and μ = 0.100 M KNO_3

8.2 Non-Aqueous Solvents. Potentiometry is not limited to aqueous solutions, even with the use of the glass electrode. The behavior of the glass electrode-calomel reference electrode system for the determination of $[H^+]$ has been described by Bates[3] for methanol, ethanol, and methanol and ethanol-water mixtures, and by Van Uitert et al.[4] for mixtures of dioxane and water up to 70% v/v dioxane. The latter solvent system was recently employed to study the equilibrium constants for the formation of cobalt(II) Schiff base complexes and their oxygenation constants.[5] The use of the low dielectric constant mixed solvent was necessary to keep all components in solution, It also provided the advantage of promoting higher degree of formation of Schiff bases, which are generally highly dissociated in aqueous solution. Low dielectric constant solvents also promote more complete formation of metal complexes.

The example given here to illustrate the method is the potentiometric equilibrium study of the salicylaldehyde, ethylenediamine, 4-methylpyridine, Co(II), dioxygen system. The method employed involved the determination of as many dissociation constants and stability constants as possible with simple combinations of the components, before working with more complex mixtures, as was illustrated above for the PXBDE system. This is apparently the first determination of stability constants and oxygenation constants for a Schiff base ligand, which is incompletely formed in solution in the absence of the metal ion, and a separate axial base. The equilibrium constants obtained are presented and species distribution curves as a function of p[H] are illustrated in Figure 8.2.

8.3 Complex Multicomponent Systems

The calculations described above for the generation of distribution curves for individual species in solution may be applied to complex systems containing large numbers of metal ions and ligands. An example has been presented in the recent paper by Motekaitis and Martell[6] describing a system contain-

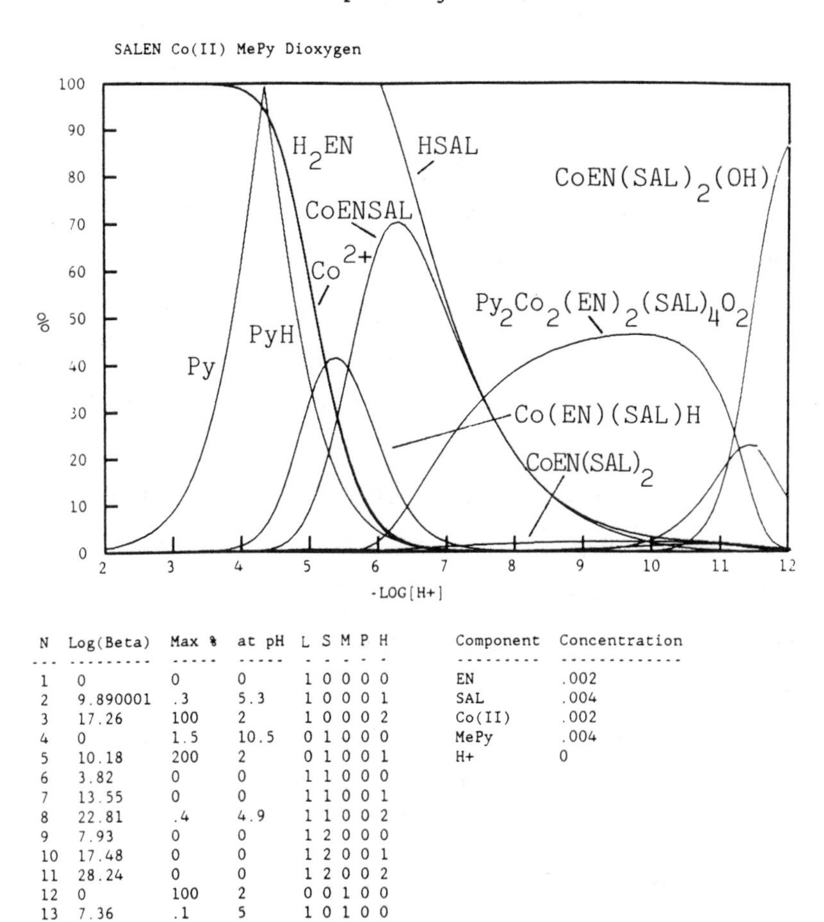

N	Log(Beta)	Max %	at pH	L	S	M	P	H
1	0	0	0	1	0	0	0	0
2	9.890001	.3	5.3	1	0	0	0	1
3	17.26	100	2	1	0	0	0	2
4	0	1.5	10.5	0	1	0	0	0
5	10.18	200	2	0	1	0	0	1
6	3.82	0	0	1	1	0	0	0
7	13.55	0	0	1	1	0	0	1
8	22.81	.4	4.9	1	1	0	0	2
9	7.93	0	0	1	2	0	0	0
10	17.48	0	0	1	2	0	0	1
11	28.24	0	0	1	2	0	0	2
12	0	100	2	0	0	1	0	0
13	7.36	.1	5	1	0	1	0	0
14	14.28	0	0	2	0	1	0	0
15	17.84	0	0	3	0	1	0	0
16	4.84	0	0	2	0	1	0	-1
17	17.03	70.2	6.3	1	1	1	0	0
18	22.6	41.5	5.4	1	1	1	0	1
19	21.48	2	9	1	2	1	0	0
20	11.54	22.6	11.4	1	2	1	0	-1
21	.42	87.2	12	1	2	1	0	-2
22	0	199.2	12	0	0	0	1	0
23	4.36	199.1	2	0	0	0	1	1
24	54.02	46.3	9.6000	2	4	2	2	0
25	0	0	0	0	0	0	0	1
26	-16	0	0	0	0	0	0	-1

Component	Concentration
EN	.002
SAL	.004
Co(II)	.002
MePy	.004
H+	0

Figure 8.2. Species distribution of Co(II)-SALEN as a function of p[H] in 70% dioxane aqueous solution at 25.0°C, μ = 0.100 M KCl saturated with dioxygen. The auxiliary base is 4-methylpyridine.

ing six metal ions, (Cu(II), Ni(II), Co(II), Zn(II), Ca(II) Fe(III)) at 1.0 x 10^{-6} M, and five multidentate ligands: citrate (CIT), tripolyphosphate (TPP), nitrilotriacetate (NTA), triaminotriethylamine (TREN), and N,N'-di(o-hydroxybenzyl)ethylenediamine-N,N'-diacetate (HBED). With the use of stability constants previously determined, computer calculations were carried out to show the equilibrium distributions of metal ions among the various ligands available as the p[H] is varied from 2.0 to 11.0. With ligand concentrations an order of magnitude higher than metal ion concentrations, there was always an excess of all of the ligands.

The computations were carried out in FORTRAN 77 with a VAX 11/780 with program SPE (see appendix) and Program MINEQL.[7] SPE is a modification of Program BEST for the calculation of species distributions from equilibrium constants. Program MINEQL was modified to adapt it to the VAX requirements for linking plotting routines to related functions, and for streamlining I/O options. At that time (1985) SPE did not possess the ability to treat solid phases in equilibrium with species in solution. However in this system no precipitates were formed and SPE was found to give shorter run times because of its more direct coding. Another disadvantage of MINEQL is the fact that the attached data bank contains stability constants taken from non-critical compilations additionally adjusted to zero ionic strength by means of the Davies equation. It was employed for the present calculations only after substitution of critical stability constants[8] for the stability constants taken from other sources.[9,10,11] The initial concentrations of metal ions and ligands investigated are given in Table 8.2 and the major complexes formed in solution are indicated in Table 8.3. The numbers of various types of species considered capable of forming in the mixture are presented in Table 8.4. The species distributions (p[H] profiles) obtained for this system at low Ca(II) concentration (1.0×10^{-6} M) and high Ca(II) concentration (1.0×10^{-2} M) are given in Figures 8.3 and 8.4. Comparison of these two sets of species p[H] profiles shows that the one difference between the two systems (high Ca^{2+} concentration in the second) has little effect on distributions of species below p[H] 5, but has a profound effect at high p[H]. The main reason for the change is substantially complete conversion of CIT, NTA, and TPP to their Ca(II) complexes above p[H] 5, and similar conversion of HBED to its Ca(II) chelate at higher p[H], thus leaving TREN as the only ligand in basic solution not complexed by Ca(II). For a more detailed discussion of the results the reader is referred to the paper by Motekaitis and Martell.[6]

Table 8.2. Concentration of Metal Ions and Ligands[a,b] Investigated.

Components		Molar Concentration	
		System I	System II
Cu(II)		1.00×10^{-6}	1.00×10^{-6}
Ni(II)		1.00×10^{-6}	1.00×10^{-6}
Co(II)		1.00×10^{-6}	1.00×10^{-6}
Zn(II)		1.00×10^{-6}	1.00×10^{-6}
Ca(II)		1.00×10^{-6}	1.00×10^{-2}
Fe(III)		1.00×10^{-6}	1.00×10^{-6}
NTA H_3L	(N)	1.00×10^{-5}	1.00×10^{-5}
TREN L	(T)	1.00×10^{-5}	1.00×10^{-5}
CIT H_3L	(C)	1.00×10^{-5}	1.00×10^{-5}
TPP H_5L	(P)	1.00×10^{-5}	1.00×10^{-5}
HBED H_4L	(B)	1.00×10^{-5}	1.00×10^{-5}

[a] Abbreviations in parenthesis are used in Figures 8.3 and 8.4. [b] NTA = nitrilotriacetic acid; TREN = triaminotriethylamine; CIT = citrate; TPP = tripolyphosphate; HBED = N,N'-di(o-hydroxybenzyl)ethylenediamine-N,N'-diacetic acid.

Table 8.3. Major Species present[a] at pH 7 and pH 8.1.

Species	System I		System II	
	pH 7.0	pH 8.1	pH 7.0	pH 8.1
CuHBED·H	95	61	94	54
CoHBED·H	83	37	88	28
NiHBED·H	63	62	73	54
ZnHBED·H	55	28	77	16
ZnNTA	29			
$NiHBEDH_2$	20		26	
NiNTA	15			
CoHBED	13	72	13	70
CaTPP		34	93	99
ZnTREN		47	10	71
NiHBED		30		25
CaNTA		28	98	100
CuHBED		27		27
ZnHBED		23		14
CaCIT			97	97
CaHBED·H				25

[a] Numbers given are the percentage of metal present in the form of the species indicated, except for Ca(II) in System II where the percentage of the ligand in the Ca(II) complex is indicated.

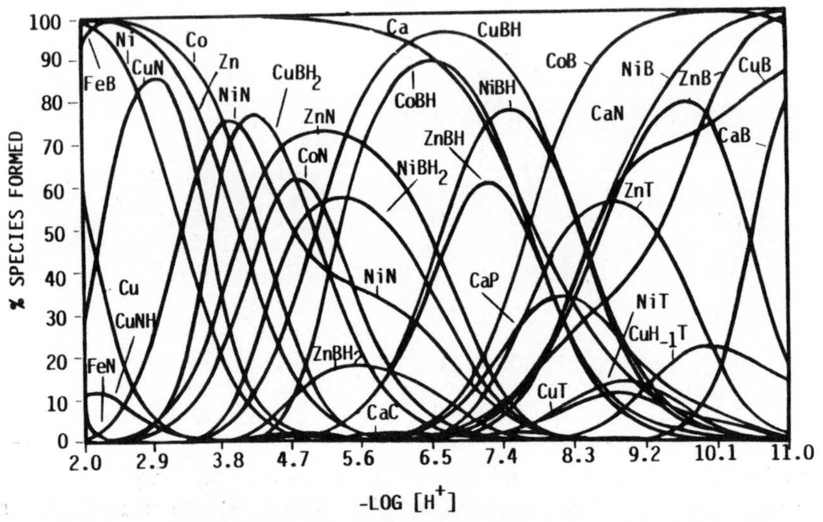

Figure 8.3. Species distribution showing only metal-containing species as a function of -log $[H^+]$. The components and their analytical concentrations are listed in Table 8.2. System I.

Table 8.4. Numbers and Types of Species that may be formed from the Components of the Systems Modeled in This Study.

Component	Type				
	Free	MH_1L	$M(OH)_jL$	Solid[a]	Total
Cu(II)	3	12	3	1	19
Ni(II)	2	11	1	1	15
Co(II)	4	11	2	1	18
Zn(II)	5	10	1	1	17
Ca(II)	2	10	0	1	13
Fe(III)	5	6	5	1	17
NTA	4	-	-	-	4
TREN	4	-	-	-	4
CIT	4	-	-	-	4
TPP	4	-	-	-	4
HBED	7	-	-	-	7
				Total	122

[a] The possibility of a precipitate was included in the calculations but none was found.

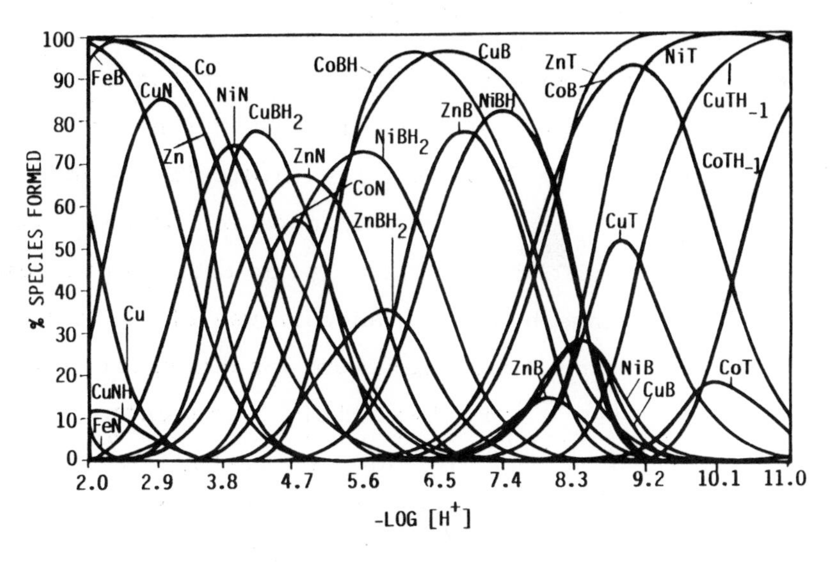

Figure 8.4. Species distribution showing only metal-containing species as a function of -log $[H^+]$. The components and their analytical concentrations are listed in Table 8.2. System II.

8.4. Equilibrium with Solid Phases

Metal ions may form insoluble hydroxides or insoluble complexes. In addition, in some p[H] regions the ligand itself may show a tendency toward separation from solution. For systems which form insoluble phases the mass balance relationships employed with homogeneous solutions for T_L, T_M and T_H do not hold. The quantities of materials removed from solution are governed by solubilities, and solubility products, K_{sp}'s.

Important K_{sp}'s frequently encountered in equilibrium work are the solubility products for the metal hydroxides of the divalent and the trivalent metal ions. Equations 8.24-8.27 describe two examples of each, respectively.

$$[Mg^{2+}][OH^-]^2 = 7.1 \times 10^{-12} \tag{8.24}$$

$$[Co^{2+}][OH^-]^2 = 2.5 \times 10^{-15} \tag{8.25}$$

$$[La^{3+}][OH^-]^3 = 2.0 \times 10^{-21} \tag{8.26}$$

$$[Fe^{3+}][OH^-]^3 = 2.5 \times 10^{-39} \tag{8.27}$$

The solubility product for Mg(II) is greater than that of Co(II), hence $Co(OH)_2$ is less soluble than $Mg(OH)_2$ at a given p[H] or alternatively it takes less hydroxide to exceed the solubility product for a given concentration of Co^{2+} relative to Mg^{2+}. Since such solubility products represent a description of a phase boundary, another way of utilizing them is to check how much (what maximum concentration) of free metal ion can exist at a given $[OH^-]$ concentration. For example in the case of Fe^{3+} let us calculate $[Fe^{3+}]$ at p[H] 2.0 where the $[OH^-]$ is approximately 1×10^{-12} M.

$$[Fe^{3+}] = 2.5 \times 10^{-39}/(1 \times 10^{-12})^3 = 2.5 \times 10^{-3} \text{ M}$$

This result is consistent with the well-known fact that Fe^{3+} precipitates very readily from stock solutions and strong acid must be present to counteract the tendency toward $Fe(OH)_3$ formation in solutions greater than millimolar in concentration. A similar calculation for La^{3+} based on equation 8.27 yields $2 \times 10^{+15}$ (p[H] 2), $2 \times 10^{+6}$ (p[H] 5) and 2×10^{-3} (p[H] 8) showing that only near p[H] 8 will millimolar La^{3+} solutions become turbid.

An important use of solubility products is the checking of the results of equilibrium measurements. The free metal ion concentration must always be less than that required to satisfy its solubility product with the $[OH^-]^n$ through the entire p[H] range. If the product is greater than the K_{sp}, then the data should be considered invalid over the range of p[H] values, at which this situation exists. However, there are cases where the solution may be stable, yet supersaturated with respect to the insoluble component. In such cases the data may be considered as being at equilibrium, and the equilibrium constants thus determined are correct, even though the system is metastable and precipitates on standing over a long period of time.

If the ligand is insoluble in water, it is not always necessary to resort to solvent modification. Since usually the insoluble form is the uncharged (non-ionic) protonated form, the ligand can usually be solubilized either in acidic or basic solutions wherein the ligand carries a positive or negative charge and is more strongly solvated. In such cases titrations may be carried out rapidly in order to take measure as far into the supersaturated region as possible.

There are occasions when the cation of the supporting electrolyte results in the formation of an insoluble salt of the ligand or complex. Sometimes, simply a change of the ionic medium is sufficient to achieve more favorable solubility properties for the system being studied.

8.5 Equilibrium Involving Hydrolytic Species

Many metal ions have tendencies toward formation of hydroxo or olated species. Some hydrolytic products are mononuclear and have simple structures and formulas, while other are polynuclear and very complex. Aluminum(III), for example, forms a monohydroxo species at low p[H] (eq. 8.28) and the aluminate anion at high pH (eq. 8.29). Some of the intermediate structures are polynuclear and while metastable under most conditions, may persist for long periods of time without reverting to more stable species, such as solid $Al(OH)_3$. An outstanding example is the complex $Al_{13}(OH)_{32}^{7+}$, which has been definitely identified and its formation constant has been reported.[12] For further information on the hydrolytic species of Al(III) the reader is referred to papers by Martell and Motekaitis[13] and Bertsch and Anderson.[14] The corresponding tetrahydroxogallium complex, $Ga(OH)_4^-$, forms readily and the equilibrium shown in equation 8.30 has been shown to be useful for the computation of strong Ga^{3+} binding constants.[15]

$$Al^{3+} + OH^- \rightleftharpoons AlOH^{2+} \qquad \text{Log } K = 8.48 \qquad (8.28)$$

$$Al^{3+} + 4OH^- \rightleftharpoons Al(OH)_4^- \qquad \text{Log } K = 33.0 \qquad (8.29)$$

$$Ga^{3+} + 4OH^- \rightleftharpoons Ga(OH)_4^- \qquad \text{Log } K = 39.4 \qquad (8.30)$$

Hydrolysis of aquo iron(III) systems have also been studied and five mono and binuclear soluble species have been identified; nevertheless in the course of potentiometric measurements such complexes are not in equilibrium, and when the ligands are weakly coordinating, eventually (sometimes after periods of time up to a week) solid, finely divided, gelatinous ferric hydroxide separates and becomes visible. Undoubtedly, phase separation of the solid takes place much earlier than the time at which it becomes visible. Hydrolysis can be considered as taking place when the investigator observes a steady drift in p[H] and preliminary calculations indicate that the calculated concentration product is on the order of magnitude of the published solubility product.

A method of dealing with solubility problems which is sometimes successful is to abandon the millimolar potentiometric methods and resort when possible to spectrophotometric methods of measuring complex equilibria. These methods have the benefit of using much lower concentrations (sometimes on the order of 10^{-5} M) and the consequent suppression of the formation of polynuclear species. An additional advantage is the economy of using very small amounts of ligand.

8.6 Species Distributions of Hydroxo and Fluoro Complexes of Aluminum(III)

An example of the use of program SPE is the determination of the hydrolytic species present in solution in competition with precipitation of solid $Al(OH)_3$.[13,15] Table 8.5 lists the equilibrium data available for fluoride and hydroxide complexes of Al(III). If one uses only the equilibrium data for hydroxo species in solution, one obtains the distribution diagram illustrated in Figure 8.5 for total metal species concentrations of 10^{-3} M. It is seen that in the intermediate p[H] range, hydrolytic complexes, including $Al_{13}(OH)_{32}^{7+}$, predominate. If one takes into account the solid phase, $Al(OH)_3$, by including its k_{sp} ($10^{-33.5}$) in the calculation, an entirely different picture emerges (Figure 8.6) in that the solid predominates over the entire p[H] range except for the extremes below p[H] 4 and above p[H] 11.5. For 10^{-6} molar metal ion, the soluble $Al(OH)_3$ complex shows up at approximately 5×10^{-8} M (5% of the total) (Figure 8.7). In the presence of 5×10^{-5} M fluoride ion, it is seen (Figure 8.8) that fluoride competes strongly with hydroxide for

Al(III) in the acid region, and pushes up the p[H] of first precipitation by two units, from ca. 4 to ca. 6.

Table 8.5. Equilibrium Constants for Hydroxo and Fluoro Al(III) Complexes

Ligand	Equilibrium Quotients and Log K Values			
OH^-	$\frac{[AlOH^{2+}]}{[Al^{3+}][OH^-]}$	8.21[a]	$\frac{[Al(OH)_2^+]}{[Al^{3+}][OH^-]^2}$	19.0[a]
	$\frac{[Al(OH)_3]}{[Al^{3+}][OH^-]^3}$	27.0[b]	$\frac{[Al(OH)_4^-]}{[Al^{3+}][OH^-]^4}$	31.4[b]
	$\frac{[Al_3(OH)_4^{5+}]}{[Al^{3+}]^3[OH^-]^4}$	41.35[a]	$\frac{[Al_{13}O_4(OH)_{24}]^{7+}}{[Al^{3+}]^{13}[OH^-]^{32}}$	33.02[a]
F^-	$\frac{[AlF^{2+}]}{[Al^{3+}][F^-]}$	6.43[b]	$\frac{[AlF_2^+]}{[AlF^{2+}][F^-]}$	5.20[b]
	$\frac{[AlF_3]}{[AlF_2^+][F^-]}$	3.9[b]	$\frac{[AlF_4^-]}{[AlF_3][F^-]}$	2.8[b]

[a] t = 25°C, 0.6 M; [b] t = 25°C, 0.1 M.

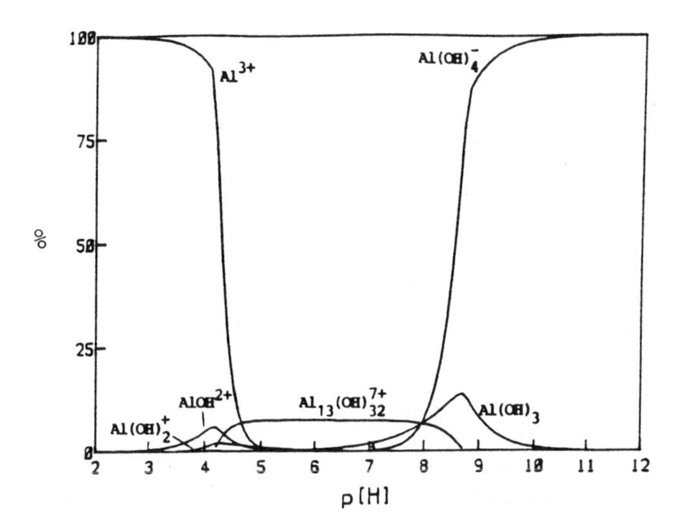

Figure 8.5. Relative molar concentration of hydrolytic species formed from aquo Al(III) ion as a function of -log $[H^+]$. Total analytical concentration of Al(III) species is 1.00 x 10^{-3}M. The 7.6 % molar concentration of $Al_{13}(OH)_{32}$ constitutes 98.8% of the Al(III) present. t = 25.0°C, μ = 0.100 M (KNO_3).

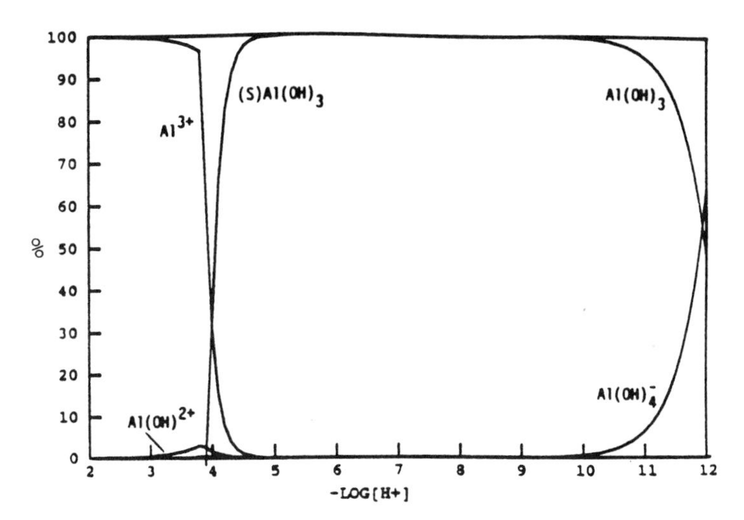

Figure 8.6. Relative molar concentration of hydrolytic species formed from aquo Al(III) ion as a function of -log $[H^+]$. Total analytical concentrations of Al(III) species if 1.00 x 10^{-3} M. $(S)Al(OH)_3$ represents the relative quantity of precipitated $Al(OH)_3$. t = 25.0°C, μ = 0.100 M (KNO_3).

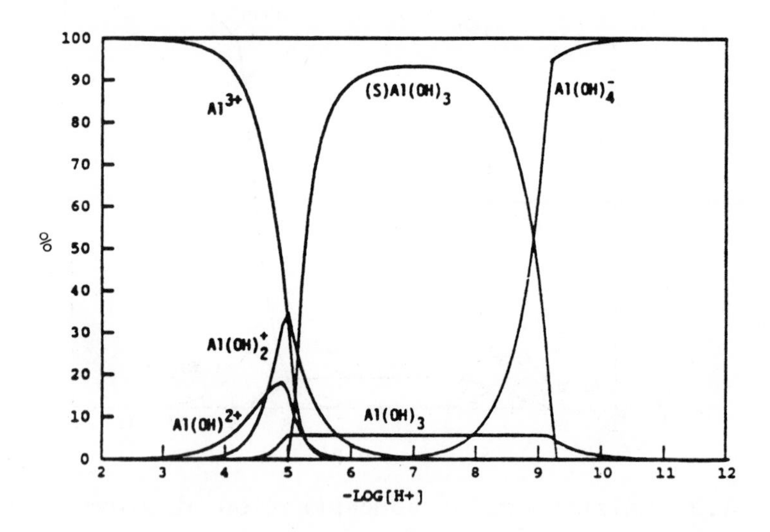

Figure 8.7. Relative molar concentration of hydrolytic species formed from aquo Al(III) ion as a function of -log $[H^+]$. Total analytical concentrations of Al(III) species if 1.00 x 10^{-6} M. (S)$Al(OH)_3$ represents the relative quantity of precipitated $Al(OH)_3$. t = 25.0°C, μ = 0.100 M (KNO_3).

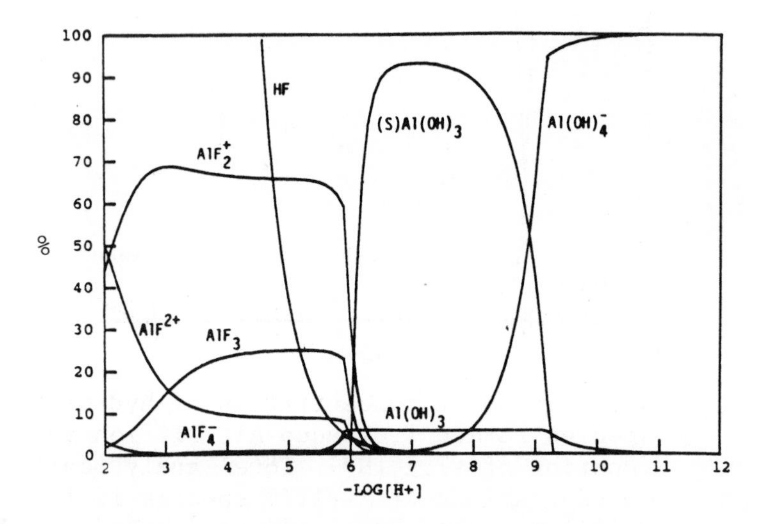

Figure 8.8. Relative molar concentration of complexes species formed from 1.0 x 10^{-6} M aquo Al(III) ion and 5.0 x 10^{-5} M fluoride ion. (S)$Al(OH)_3$

References

1. Ng, C. Y.; Motekaitis, R. J.; Martell, A. E. Inorg. Chem. **1979**, 18, 2982.
2. Ng, C. Y.; Martell, A. E.; Motekaitis, R. J. J. Coord. Chem., **1979**, 9, 255.
3. Bates, R. G. "Determination of pH"; Wiley: New York, 1973.
4. Van Uitert, L. G.; Fernelius, W. C. J. Am. Chem. Soc., **1953**, 75, 451.
5. Motekaitis, R. J.; Martell, A. E.; Nelson, D. A. Inorg. Chem., **1984**, 23, 275.
6. Motekaitis, R. J.; Martell, A. E. J. Coord. Chem., **1985**, 24, 49.
7. J. C. Westall, J. L. Zachary, F. M. M. Morel, Technical Note No.18, EPA Grant No.R-803738, July, 1976.
8. Smith, R. M.; Martell, A. E. "Critical Stability Constants" Vol.1, 2, 3, 4, 5; Plenum: New York, 1974, 1975, 1977, 1976, 1982 (Vol.6 to by published early 1989).
9. "Stability Constants, Part II, Organic Ligands" Special Publication No.17, Eds. Sillen, L. G.; Martell, A. E.; Chemical Society: London, 1964.
10. "Stability Constants, Supplement No.1", Special Publication No.25, Eds. Sillen, L. G.; Martell, A. E.; Chemical Society: London, 1971.
11. "Stability Constants of Metal-ion Complexes. Part B. Organic Ligands", Ed. D. D. Perrin; Pergamon Press: Oxford, 1979.
12. Mesmer, R. E.; Baes, C. F., Jr. Inorg. Chem., **1971**, 10, 2290; Ref. 8, Vol. 4.
13. Martell, A. E.; Motekaitis, R. J. Paper No.50, Abstracts of the 194th National Meeting of the American Chemical Society, Division of Environmental Chemistry, New Orleans, LA, Aug.30-Sept.4, 1987 (full paper in press).
14. Bertsch, P. M.; Anderson, M. A. Paper No.51, Abstracts of the 194th National Meeting of the American Chemical Society, Division of Environmental Chemistry, New Orleans, LA, Aug.30-Sept.4, 1987 (full paper in press).
15. Motekaitis, R. J.; Martell, A. E. Inorg. Chem., **1980**, 19, 1646.

9

CRITICAL STABILITY CONSTANTS AND THEIR SELECTION

9.1 General Criteria

The wide variations frequently found in published constants for a metal complex indicate the presence of errors whose sources include impure ligand, poor experimental design, faulty experimental conditions, inaccurate measurements, and erroneous calculations (see Chapter 4). Although it has been suggested that decisions between published sets of data must sometimes be arbitrary and therefore possibly unfair,[1] we have found that the application of reasonable guidelines leads directly to the elimination of a considerable fraction of the published data of doubtful value. The following is an account of the various criteria that have been employed by this research group for the selection of data that have been published in "Critical Stability Constants".[2]

Lack of Essential Experimental Control. The reliability of the scientific data is revealed frequently by the care with which the reaction conditions and measurements are controlled and described. Papers deficient in specifying essential reaction conditions (i.e., temperature, ionic strength, nature of supporting electrolyte) are excluded from the compilation. Also used as a basis for disqualification of published data is lack of information on the purity of the ligand. Two common deficiencies are poor calibration of the potentiometric apparatus and the failure to define the equilibrium quotients reported in the paper. Papers in which temperature or ionic strength are not controlled are omitted from the compilation.

Selection of a Constant from a Number of Reported Values. When several workers are in close agreement on a particular constants the average of their results is selected.

Values showing considerable deviation are eliminated. In cases where the agreement is poor and few results are available for comparison, more subtle methods are needed to select the best value. The selection may be guided by a comparison with values obtained for other metal ions with a ligand and with values obtained from the same metal ion with similar ligands. While established trends among similar metal ions and among similar ligands are valuable in deciding between widely varying data, such guidelines are used cautiously, so as not to overlook occasional real examples of specificity or anomalous behavior.

In some cases the constants reported by several workers for a given group of metal ions would have similar relative values, but would differ considerably in the absolute magnitudes of the constants. Then a set of values from one worker near the median of all values reported is selected as the best constants. By this method it is believed that internal consistency is preserved to a greater extent than would be obtained by averaging reported values for each individual metal ion.

Reliability of Investigator. When there is poor agreement between published values, and comparison with other metal ions and ligands does not suggest the best value, the results of more experienced research groups who have supplied reliable values for other ligands are selected. When such assurances are lacking it is sometimes possible to give preference to values reported by an investigator who has published other demonstrably reliable values obtained by the same experimental methods.

Single Investigator. Values reported by only one investigator are included in the data base unless there is some reason to doubt their validity. It is recognized that some of these values may be in error, and that such errors will probably not be detected until the work is repeated by other investigators, or until more data become available for analogous ligands or other closely related metal ions. Some values involving unusual metal ions or unusual ligands may be omitted if there are serious questions about the nature of the complexes formed.

Bibliography. A complete bibliography for each ligand is included so that the user of the data base may determine the completeness of the literature search employed in the selection of critical values. The user may also employ these references to make his own evaluation if he has any questions or reservations concerning the selection.

The references cited consist of two groups, those that were used for the selection of the critical constants, and those that were not. The inclusion of the latter group is considered important to assure the user about the completeness of the literature search, as well as to provide references for any supplementary studies that the user may wish to undertake.

An important factor in the organization of the bibliography is the grouping of all pertinent references employed in selecting a particular equilibrium constant. A critical constant listed in the tables is never tied to a particular reference, unless, of course, there is only one reference for that quantity. Thus if a user is skeptical about the choice of a constant, he would have to study all the cited references and draw his own conclusions. By going through the same procedure as that followed by the compilers, the user may see somewhat more clearly the reasonableness of the choice that was made.

9.2 Examples of Critical Data Selection

The complexity of the problems involved in the selection of critical stability constants is best illustrated by comparing tables of the literature constants for representative metal ions and ligands with tables of the corresponding critical constants derived from them, and examining the procedure employed in converting the former to the latter. EDTA is chosen as a representative ligand because of its importance and the fact that it has been widely investigated. However, it lacks one of the characteristics frequently encountered in literature studies - the problem of ligand purification and characterization - because it is readily available in very pure form. Two basic metal ions, Ca^{2+} and Mg^{2+}, and two common less basic metal ions, Cu^{2+} and Zn^{2+}, are selected for comparison. The critical protonation and stability constants, and enthalpies of reaction for EDTA complexes of these metal ions are taken from "Critical Stability Constants" Volumes 1 and 5, respectively. These data represent a complete literature survey up to and including publications appearing in 1979.

The non-critical compilation contains many numbers which in some cases vary considerably in value, and which must be reduced to a single number. The number of non-critical constants found in the literature for ionic strength near 0.1 is so extensive that only the values of K_1^H and K_1^{Cu} (eq. (9.1) and (9.2)), together with the corresponding enthalpy changes are presented in Table 9.1.

The critical values derived from the data in Table 9.1 are presented in Table 9.2 along with constants for other protonation reactions and other metal ions, 0.100 M and 25.0°C, respectively. The values of ΔH° make it possible to calculate stability constants at temperatures which are not greatly different from the temperatures originally cited. The large reduction in the numbers in the conversion of the data in Table 9.1 to the Critical Table (Table 9.2) illustrates the advantage of the critical constants to the user, especially in

Table 9.1. First Protonation Constant and Cu(II) Stability Constants of EDTA

Protonation Constants			
t,°C	Ionic Strength	Supporting Electrolyte	Log K_1^H
25	0.1	Pr_4NBr	10.39
25	0.1	Et_4NBr	10.37
25	0.1	Me_4NCl	10.39
20	0.1	Me_4NCl	10.44
25	0.1	Me_4NBr	10.35
37	0.15	Et_4Ni	10.14
30	0.1	KCl	10.25
25	0.1	KCl	10.30
20	0.1	KCl	10.32
25	0.1	KNO_3	10.27
25	0.1	KNO_3	10.25
25	0.1	KNO_3	10.34
25	0.1	KNO_3	10.21
25	0.1	KNO_3	10.20
20	0.1	KCl	10.26
30	0.1	KNO_3	10.12
22	0.1	KNO_3	10.23
25	0.1	KNO_3	10.18
20	0.1	KCl	10.23
20	0.1	KNO_3	10.23
25	0.1	KNO_3	10.15
25	0.1	KCl	10.11
25	0.1	KNO_3	10.2
35	0.15	KCl	10.05
25	0.1	KNO_3	10.07
37	0.15	$NaClO_4$	9.74
25	0.1	$NaNO_3$	9.50
37	0.15	NaCl	9.14
37	0.15	NaCl	9.12

Enthalpies of Protonation			
t,°C	Ionic Strength	Supporting Electrolyte	$\Delta H°_1^H$
20	0.1	KNO_3	-6.0
20	0.1	Me_4NCl	-5.9
25	0.1	KNO_3	-5.8
20	0.1	KCl	-5.69
20	0.1	KNO_3	-5.67
25	0.1	KNO_3	-5.59
25	0.1	KNO_3	-5.5
25	0	KCl	-5.3

Cu(II) Stability Constants			
t,°C	Ionic Strength	Supporting Electrolyte	Log K_1^{Cu}
25	0.1	KNO_3	19.19
5	0.1	KNO_3	19.00
20	0.1	KNO_3	18.80
25	0.1	KNO_3	18.83
20	0.1	$KClO_4$	18.92
25	0.1	KNO_3	18.80
30	0.1	$KClO_4$	17.8
20	0.1	KCl	18.86
25	0.1	KNO_3	18.71
30	0.1	KNO_3	17.7
25	0.1	$NaNO_3$	18.7
25	0.1	$NaClO_4$	18.7
20	0.1	KCl	18.38

Enthalpies of Cu(II) Complex Formation			
t,°C	Ionic Strength	Supporting Electrolyte	$\Delta H°_1^{Cu}$
25	0.1	KNO_3	-8.3
25	0.1	KNO_3	-8.8
20	0.17	KNO_3	-8.67
25	0.2	$NaNO_3$	-8.2
20-25		corr for Na^+	-8.3
25	0.06	Na^+	-8.2
20	0.1	KNO_3	-8.15

view of the fact that the constants in Table 9.2 may be considered more reliable. For the purposes of discussion, two reactions are singled out for close scrutiny, the first protonation constant of EDTA and the formation of the 1:1 Cu(II) complex:

$$H^+ + L^{4-} \rightleftharpoons HL^{3-} \qquad K_1^H = \frac{[HL^{3-}]}{[L^{4-}][H^+]} \qquad (9.1)$$

$$Cu^{2+} + L^{4-} \rightleftharpoons CuL^{2-} \qquad K_1^{Cu} = \frac{[CuL^{2-}]}{[L^{4-}][Cu^{2+}]} \qquad (9.2)$$

Table 9.2. Selected Critical Stability Constants for EDTA at 25.0°C and 0.1 M Ionic Strength

Metal Ion	Equilibrium Quotient	Log K with R_4N^+	Log K with K^+	ΔH°
H^+	HL/H·L	*10.37 ± 0.02	*10.19 ± 0.04	*-5.6 ± 0.2
	H_2L/H·HL	6.13 ± 0.03	6.13 ± 0.03	-4.2 ± 0.2
	H_3L/H·H_2L	2.68	2.69 ± 0.05	+1.4 ± 0.1
	H_4L/H·H_3L	1.96	1.99 ± 0.1	+0.3 ± 0.0
	H_5L/H·H_4L		1.5 ± 0.1	+0.6
Mg^{2+}	ML/M·L	9.13	8.85 ± 0.08	+3.3 ± 0.2
Ca^{2+}	ML/M·L	10.97	10.65 ± 0.08	-6.5 ± 0.1
Cu^{2+}	ML/M·L		*18.78 ± 0.08	*-8.2 ± 0.1
Zn^{2+}	ML/M·L		16.5 ± 0.1	-4.7 ± 0.2

* Constants derived from data in Table 9.1.

The first protonation constant of EDTA varies with background electrolyte because of weak complex formation by the tetranegative anion of EDTA, L^{4-}, with Na^+ and K^+. The values reported for tetraalkylammonium salts as supporting electrolyte show a very small change as the alkyl group is varied, but this change has about the same magnitude as the uncertainties in the measurements, and the one value selected for R_4N^+ supporting electrolyte is all that is justified. The value at 0.15 ionic strength is slightly low and therefore not considered. The average of the other five values, 10.37 with a range of ±0.02 is selected for tetraalkylammonium media. With K^+ media as the supporting electrolyte the average of nine of eighteen values, 10.19 ± 0.04, is selected. The values listed which are lower than the selected value may be low because of the use of Na_2H_2EDTA as the form of the ligand employed. All of the values in Na^+ media are much lower than values in K^+ media because of the relative magnitudes of the stability constants of KEDTA (log K = 0.8) and NaEDTA (log K = 1.8). Therefore the best value of the first protonation constant is the one given for tetraalkylammonium supporting electrolyte. Measurements in K^+ and Na^+ supporting electrolytes for the determination of the stability constants of EDTA with other metal ions should be corrected for K^+ or Na^+ binding, as the case may be. All of the data necessary for such calculations are given in the critical stability constant tables.

In order to select the best value of a particular stability constant at a specified temperature, data for that constant at other temperatures may be utilized and corrected to the desired temperature when enthalpy changes are available for the reaction under consideration. Such corrected constants are frequently helpful when other data are incomplete, but are not given the same weight as are the measurements made at the standard temperature being reported.

For the Cu^{2+} stability constants, the median value of eleven of the thirteen reported constants is selected as 18.78 ± 0.09, for a 0.10 M supporting electrolyte containing K^+. The stability constant (log K = 0.8) may be used for correcting this constant to the least complexing supporting electrolyte, containing Me_4N^+ as the cation. For ΔH°, three values at -8.8

± 0.2 kcal/mole and four values at -8.2 ± 0.1 kcal/mole are obtained after adjusting all values to 25.0°C with the variation with temperature provided by the first entry. The greater number of measurements and better agreement for the lower value leads to the selection of -8.2 kcal/mole as the most reliable constant.

The spread of reported (non-critical) constants of EDTA is much lower than the values reported for many other ligands. A more typical example of the differences between the critical and non-critical compilations is presented in Tables 9.3 and 9.4 for Cu(II)-DTPA at 0.10 ionic strength. The constants listed in non-critical Table 9.3 contain stability constants that differ by as much as one and a half log units. While such tables may be of some value as literature surveys, they are useless for the non-specialist who needs the data for work in his field, or for teaching purposes. Critical values of the type indicated in Table 9.4, on the other hand, provide the data needed in a convenient form.

9.3 Need for Additional Critical Constants

There is a growing body of data in the literature that deal with metal complex formation in non-aqueous solvents, and in mixed solvent systems. Stability constants of the complexes formed in non-aqueous solvents are important for the understanding of catalysis by metal ions, metal separations by solvent extraction, and the mechanism of metal ion transport through lipid-type membranes. Selection of critical stability data for such systems will probably be an important objective for the future development of critical stability constants.

Table 9.3. Complete List of Stability Constants for Cu^{2+} + DTPA

Method*	Temperature	Ionic Strength	Background Electrolyte	Reported Log K	Ref.
ise	25°	0.10	KNO_3	21.45	3
Hg	20°	0.10	$NaNO_3$	21.6	4
gl	20°	0.10	KCl	21.53	4
gl	25°	0.10	KNO_3	21.1	5
gl	20°	0.10	KCl	21.03	6
Hg	25°	0.10	$NaClO_4$	20.5	7
gl	20°	0.10	NaOH	20.1	4

* ise = ion selective electrode; Hg = mercury electrode; gl = glass electrode with competition from another ligand as pH is varied.

Table 9.4. Selected Critical Stability Constant for Cu^{2+} + DTPA at 25.0°C and 0.1 M ionic Strength

Cu^{2+}	ML/M·L	Log K	21.40 ± 0.05

References

1. Sillen, L. G., private communication, 1964.
2. Smith, R. M.; Martell, A. E. "Critical Stability Constants", Vol. 1, 2, 3, 4, 5; Plenum: New York, 1974, 1975, 1977, 1976, 1982. (Vol.6 to be published early 1989).
3. E. W. Baumann, J. Inorg. Nucl. Chem., **36**, 1827 (1974).
4. G. Anderegg, P. Nageli, F. Muller and G. Schwarzenbach, Helv. Chim. Acta, **42**, 827 (1959);
5. S. Chaberek, A. E. Frost, M. A. Doran and N. J. Bickwell, J. Inorg. Nucl. Chem., **11**, 184 (1959);
6. E. J. Durham and D. P. Ryskiewick, J. Am. Chem. Soc., **80**, 4812 (1958);
7. E. Wanninen, Acta Acad. Aboensis, Math. et Phys., **21**, No.17 (1960);

10

DEVELOPMENT OF A COMPLETE CRITICAL METAL COMPLEX DATA BASE

10.1 Introduction

In spite of the large number of protonation constants and metal complex stability constants that have been published, the measured constants constitute only a small fraction of the possible combinations of ligands and metal ions that are actually formed or potentially formed in biological and environmental systems. It is estimated that the number of critical constants available constitute only about 5% of the total number of constants that would be needed for a complete description of metal complex speciation in environmental and biological systems. However, the experimental determination of equilibrium constants involving twenty to thirty metal ions and thousands of organic and inorganic ligands is virtually impossible. For that reason it is necessary to estimate equilibrium constants for those complexes which have not and probably will not be measured. As will be shown below the procedures recommended for accomplishing this involve the use of the critical constants already compiled as guidelines for the estimation of the constants of complexes that have not been determined. The five steps needed for calculation of the complex species present in environmental systems containing a wide variety of organic ligands as well as a large number of metal ions are:

1. Selection of critical stability constants from published equilibrium data.
2. Evaluation of the procedures available for correlating and predicting stability constants of metal complexes.
3. Identification of the types of ligands and the metal ions for which more equilibrium data are needed for

determining metal ion speciation in multicomponent systems of interest.

4. Estimation of unknown stability constants needed for the development of the expanded data base required for calculation of metal speciation.

5. Calculation of metal complex distribution in selected systems of environmental interest, and changes in metal speciation that occur when the systems are perturbed by the addition of complexing ligands and additional metals through environmental pollution.

Similar steps, with appropriate variations in 3, 4, and 5, may be employed for complex biological systems.

The most complete compilation of critical constants is that of Martell and Smith.[1] It was conceived some fifteen to twenty five years ago as an authoritative reference and was an outgrowth of the extensive, non-critical compilations by Sillen and Martell[2,3] sponsored by the Commission on Equilibrium Data of the Analytical Division of IUPAC. It is a five volume collection with the sixth in press. With these compilations an investigator can consult a single source and find the constant(s) of interest without the problem of having to select a number from among the various published, often scattered, and frequently incorrect values in the literature. In compiling these critical values each entry was scrutinized in various ways and its appearance represents the best value available. Complete literature references for all constants are listed (including references for discarded numbers) so that the users may undertake their own selection process, if they so choose.

10.2 Linear Stability Constant Correlations

There have been several attempts to develop empirical or semi-empirical relationships to predict the stability constants of complexes on the basis of the nature of the metal ion involved and the structure of the ligand. Most of the relationships are of limited applicability; perhaps the most successful is the relationship reported by Harris,[4] illustrated below as equation (10.1).

$$\text{Log } K_{ML} = \sum_{i=1}^{n} n_i X_i + [r_5 + \sum_{i=2}^{n_5} (r_5 f_5^{i-1})]_{n_5 \neq 0}$$

$$+ [r_6 + \sum_{i=2}^{n_6} (r_6 f_6^{i-1})]_{n_6 \neq 0} \qquad (10.1)$$

<u>1st term</u> stability increments for functional groups
<u>2nd term</u> stability increments for 5-membered rings
<u>3rd term</u> stability increments for 6-membered rings
<u>Predictability</u>: within 0.8 log units

Constants for Harris Equation

		Ni^{2+}	Zn^{2+}	Cd^{2+}	Fe^{3+}
X_i	amine	3.0	2.5	2.6	4.7
	carboxylate	2.1	1.6	1.87	3.4
	imidazole	2.6	1.7		
	pyridyl	2.7	1.5		
	thioether		-1.6	-0.1	
	phenol				9.4
r_5		0.8	0.9	0.9	0.4
f_5		1.0	1.1	2.4	1.0
r_6		-0.3	-0.2	-0.9	
f_6		0.9	1.7	1.4	

Predictions of equation (10.1) for a large number of nickel complexes are compared with measured values in Figure 10.1. While the correlation obtained seems impressive, close inspection indicates a spread of data of nearly an order of magnitude between predicted and measured constants, although in some cases the values are somewhat better. The deviations, however, become much greater when the structure of the ligand is considerably changed or when substitution of inert groups on the ligands produce secondary electronic or steric effects. For these reasons the use of empirical relationships at this stage is considered too limited to be employed in the development of a broad-base compilation of stability constants.

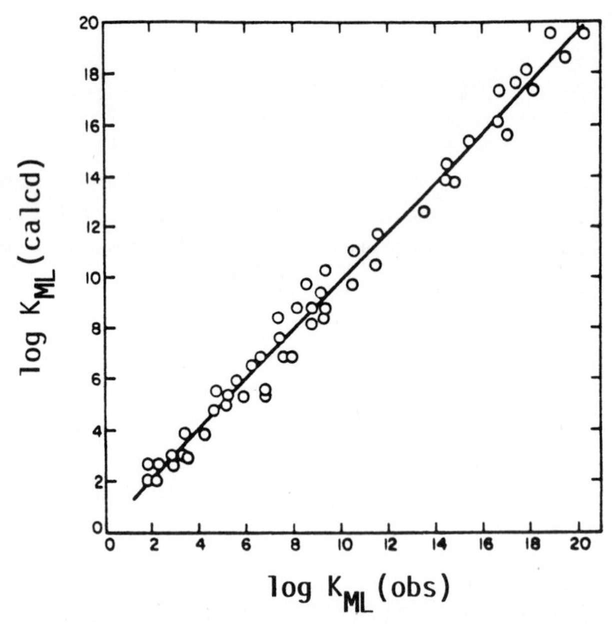

Figure 10.1. Plot of stability constants of Ni(II) complexes calculated with equation (10.1) vs. measured stability constants of the same complexes.

Careful inspection of stability constant data compiled in the critical tables[1] indicates that many of the reliable constants may be subdivided into groups of related ligands which show very little variation in the stability constants involved. The sample data listed in Table 10.1 indicate that several groups of related ligands differ in stability constants or protonation constants by as little as one tenth of a log unit. For such systems other members of that group of ligands can be assigned the same constant with confidence. For those types of ligands which show greater variation closer approximations may be made if the effect of variations in structure on stability constants are studied and the observed trends are taken into consideration. This process is greatly aided by the use of structure-stability constant relationships which have been described in the literature in the form of linear correlations. Several of the most notable of these correlations have been compiled in Table 10.2. Examples of two of these correlations are presented in Figures 10.2 and 10.3. Figure 10.2 shows that the ratio of stability constants for Ni(II) and Cu(II) complexes is nearly constant for a large number of ligands while Figure 10.3 shows a characteristic trend in stability constants observed for a large number of ligands for divalent metals in the first transition series. Such correlations and other linear relationships indicated in Table 10.2 may be used for estimating stability constants by comparing a ligand with similar ligands for a given metal ion, or if the ligand remains the same, for estimating the stability constant with a metal ion by comparison with stability constants or other metal ions recognized as showing similar behavior.

Table 10.1. Critical Protonation and Formation Constants

Ligand	Log Protonation Constant		Log Formation Constants, Cu(II)		Log Formation Constants, Ag(I)	
	NR_3	RCO_2	K_1	β_2	K_1	β_2
Primary amine	10.6±0.1		(4)		3.5±0.1	7.4±0.[illegible]
Carboxylic acid		4.7±0.1	1.8±0.1	3.2±0.5	0.7	0.6
2-Amino acid	9.6±0.1	2.3±0.1	8.1±0.1	14.9±0.1	3.3±0.3	6.7±0.[illegible]
2-Hydroxyamine	9.6±0.1		5.7	9.8	3.1±0.1	6.7±0.[illegible]
$(Glycyl)_n$amide	7.9±0.1		5.0±0.2	5.2±0.2[b]		
Dipeptide	8.1±0.1	3.1±0.1	5.5±0.3	4.2±0.7[b]		
Secondary amine	11.1±0.2				3.5±0.5	7.0±0.[illegible]
Tertiary amine	10.5±0.4				3.8±0.3	4.5±0.[illegible]

[a] 25.0°C, 0.100 M ionic strength. [b] Amide protonation constants.

Table 10.2. Examples of Early Linear Free Energy Relationships

Authors	Relationships	Ref.
Larsson (1934)	$\log K_{ML}$ ligand basicity	5
Bjerrum (1950)	$\log K_{ML}/\log K_{HL}$ = constant	6
Calvin, Wilson (1945)	$\log K_{ML}$ vs. $\log K_{HL}$ (graphical)	7
Calvin, Melchior (1948)	$\log K_{ML}$ vs. ionization potential (graphical)	8
Irving, Williams (1948)	$\log K_{ML}$ vs. atomic number	9
(1953)	(graphical)	10
Davies (1951)	$\log K_{ML}$ vs. e^2/r	11
Martell, Calvin (1952)	$\log K_{ML}$ vs. e^2/r	12
Van Uitert, Fernelius, Douglas (1953)	$\log K_{ML}$ vs. electronegativity (graphical)	13
Irving, Rosotti (1956)	$\log K_{ML} = a \log K_{HL} + b$	14
	$\log K_{ML}$ vs. $\log K_{ML'}$ (graphical)	
	$\log K_{ML}$ vs. $\log K_{M'L}$ (graphical)	
Nieboer, McBryde (1970)	$\log K_{ML} = B \log K_{M'L} + (\log K_{ML'} - B \log K_{M'L'})$	15
	$\log K_{ML} = C \log K_{ML'} + (\log K_{M'L} - C \log K_{M'L'})$	

10.3 Estimation of Stability Constants not Measured Experimentally

Examples of how such an empirical method of estimation of stability constants might work is indicated in Table 10.3. The predicted stability constants were estimated by the use of stability constants of copper complexes of analogous ligands compiled in Volumes 1 to 3 of Critical Stability Constants.[1] The measured values were not available at that time but appeared after several years in volume 5 of Critical Stability Constants, which covered a later period of time. The agreement

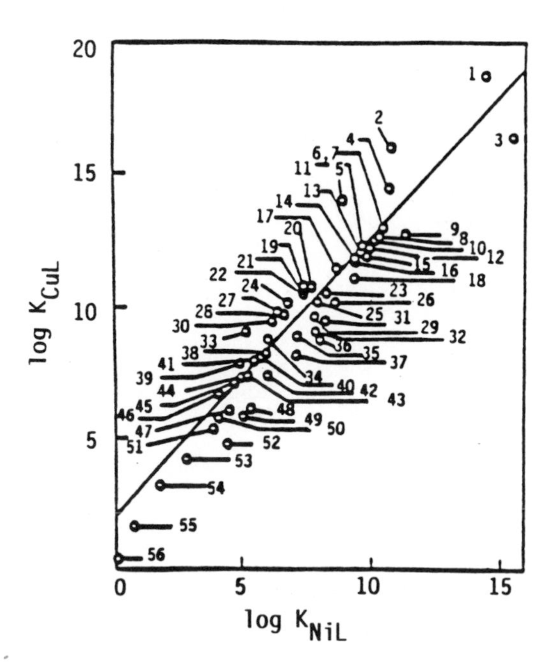

Figure 10.2.

Log K_{CuL} as a function of log K_{NiL} for complexes of a wide variety of ligands: **1**, β,β',β''-triaminotriethylamine; **2**, diethylenetriamine; **3**, ethylenediaminediacetic-dipropionic acid; **4**, Tiron (pyrocatechol-3,5-disulfonic acid); **5**, 1,2,3-triaminopropane; **6**, 8-hydroxyquinoline; **7**, dibenzoylmethane; **8**, thenoylbenzoylmethane; **9**, nitrilotriacetic acid; **10**, benzoylfuroylmethane; **11**, $CH_3COCH_2COCH_2CH_2Si(CH_3)_3$; **12**, dithenoylmethane; **13**, acetylbenzoylmethane; **14**, acetylacetone; **15**, furoylthenoylmethane; **16**, acetylthenoylmethane; **17**, $(CH_3)_3SiC_2H_4COCH_2COC_2H_4Si(CH_3)_3$; **18**, 1,2,3-triaminopropane; **19**, N-methylethylenediamine; **20**, ethylenediamine; **21**, 1,2-diaminopropane; **22**, iminopropionicacetic acid; **23**, iminodiacetic acid; **24**, N-ethylethylenediamine; **25**, 8-hydroxy-2,4-dimethylquinazoline; **26**, 8-hydroxy-2-methylquinoline; **27**, 1,3-diaminopropane; **28**, sym. N,N′-dimethylethylenediamine; **29**, 5-hydroxyquinoxaline; **30**, iminodipropionic acid; **31**, 8-hydroxycinnoline; **32**, folic acid; **33**, N-propylethylenediamine; **34**, triethylenetetramine; **35**, aspartic acid; **36**, 1,10-phenanthroline; **37**, 2,2′-dipyridyl; **38**, α-alanine; **39**, leucine; **40**, glycine; **41**, isonicotinic hydrazide; **42**, 2-methyl-1,10-phenanthroline; **43**, salicylaldehyde; **44**, sarcosine; **45**, β-alanine; **46**, riboflavin; **47**, glycylglycine; **48**, oxalate ion; **49**, 2,9-dimethyl-1,10-phenanthroline; **50**, malonic acid; **51**, salicylaldehyde-5-sulphonic acid; **52**, 4-hydroxypteridine; **53**, ammonia; **54**, pyridine; **55**, acetic acid; **56**, nitroacetic acid. The full line is of unit slope.

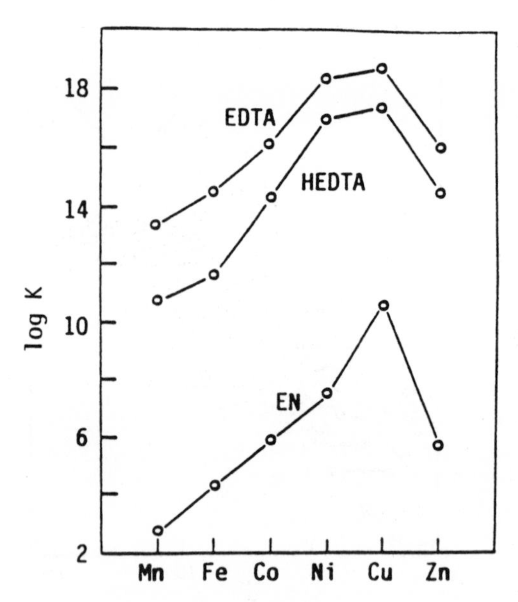

Figure 10.3. Variation of stability constants of ethylenediaminetetraacetic acid (EDTA), N-hydroxyethylenediamine,N,N',N'-triacetic acid (HEDTA), and ethylenediamine (EN) chelates with atomic number of divalent transition metal ions.

between predicted and measured values seems to be satisfactory and generally the estimated values can be considered to be accurate to about one tenth of a log unit, which is sufficient for the estimation of speciation in environmental systems. The first attempts toward expanding the critical stability constant data base to cover unmeasured values is now being carried out by the author and coworkers.[16,17,18]

Table 10.3. Predicted and Measured Stability Constants for Selected Cu(II) Complexes

Ligand	Log Stability Constants		Difference
	Predicted	Measured	
glycylglycylglycylamide	5.0 ± 0.2	4.77	-0.2
D-allo-isoleucine	8.1 ± 0.1	8.09	0.0
glycyl-L-threonine	5.5 ± 0.3	5.57	+0.1

After the development of the expanded data base of critical and estimated constants at a given ionic strength and temperature such as 0.1 M ionic strength and 25°C, it may be necessary for certain purposes to correct the values for other temperatures (e.g., physiological temperatures) and other ionic strengths (e.g., the ionic strength of seawater). Temperature corrections can be made if accurate values of the heat of reaction have been measured. If not, heats of reaction for the formation of similar complexes could be used for estimated values. Corrections for ionic strength from 0.1 to 0.7 for seawater is more difficult and theoretical relationships based on the Debye-Huckel theory or the Davies equation are completely unsatisfactory since they apply only to 1:1 electrolytes and to very low ionic strength. For corrections of this nature an empirical tabulation has been developed by the author and his coworkers[16] and is presented in Table 10.4. The numbers without brackets in this table were taken directly from selected constants available in the literature at the ionic strengths indicated. The numbers in brackets are correction factors estimated on the basis of trends in the measured values.

10.4 Metal Speciation in Sea Water With and Without Added Ligands

The following example of the evaluation of metal complex speciation in seawater, with and without an added chelating agent, is described as an example of the use of critical stability constants, an expanded data base involving estimated constants, and the adjustment of stability constants to satisfy the conditions of the system under consideration, for the determination of metal complexes present in complex environmental systems. The first attempt to determine the speciation of inorganic complexes in the oceans was published

Table 10.4. Variation of Stability Constants with Ionic Strength and Charge Relative to 0.1 M Ionic Strength. Values in [] are based on trends.

Ionic Strength		0.0	0.5	1.0	2.0	3.0
				M^{+}		
L^{-}	log K_1	+0.2	-0.1	-0.1	0.0	+0.2
	log β_2	+0.3	-0.1	0.0	+0.3	+0.6
L^{2-}	log K_1	+0.4	-0.2	-0.2	-0.1	0.0
	log β_2	+0.6	-0.4	-0.4	-0.3	0.0
L^{3-}	log K_1	+0.6	-0.3	-0.3	-0.3	-0.2
	log β_2	+1.0	-0.5	-0.5	[-0.4]	[-0.2]
				M^{2+}		
L^{-}	log K_1	+0.4	-0.2	-0.2	-0.1	0.0
	log β_2	+0.6	-0.4	-0.4	-0.3	0.0
L^{2-}	log K_1	+0.8	-0.4	-0.4	-0.4	[-0.3]
	log β_2	+1.2	-0.8	-0.8	[-0.7]	[-0.5]
L^{3-}	log K_1	+1.2	-0.6	-0.6	[-0.7]	[-0.6]
	log β_2	+1.8	[-1.0]	[-1.0]	[-1.0]	[-0.9]
				M^{+3}		
L^{-}	log K_1	+0.6	-0.3	-0.3	0.3	-0.2
	log β_2	+1.0	-0.5	-0.5	[-0.4]	[-0.2]
L^{2-}	log K_1	+1.2	-0.6	-0.6	[-0.7]	[-0.6]
	log β_2	+1.8	[-1.0]	[-1.0]	[-1.0]	[-0.9]
L^{3-}	log K_1	[+1.8]	[-0.9]	-0.9	[-1.1]	[-1.0]
	log β_2	[+2.6]	[-1.5]	[-1.5]	[-1.6]	[-1.5]
				M^{4+}		
L^{-}	log K_1	[+0.8]	[-0.4]	[-0.4]	[-0.5]	[-0.4]
	log β_2	[+1.6]	[-0.6]	[-0.6]	[-0.6]	[-0.5]
L^{2-}	log K_1	[+1.6]	[-0.8]	[-0.8]	[-1.0]	[-0.9]
	log β_2	[2.4]	[-1.4]	[-1.4]	[-1.5]	[-1.4]
L^{3-}	log K_1	[2.4]	[-1.2]	[-1.2]	[-1.5]	[-1.4]
	log β_2	[+3.4]	[-2.0]	[-2.0]	[-2.2]	[-2.1]

by Sillen some twenty years ago[19] but the calculations were hampered by incomplete data, ionic strength problems and the lack of modern computer facilities, as well as many uncertainities concerning the stability constants themselves. In the present study, described in detail elsewhere,[20] more accurate equilibrium data are now available as well as good computer facilities and methods for the elucidation of complex multicomponent systems. In order to carry out these calculations a baseline computation was set up for seawater speciation at p[H] 8.1 for the complexes present in the open ocean. A calculation was then carried out to describe the perturbation of the natural complex system by the addition of a strong chelating agent as a pollutant. The chelating agent, EDTA, was selected because it is produced and used in large quantities throughout the world, is only slowly biodegradable, and has the potential of being released to the environment in various locations.

The inorganic composition of seawater is taken from a paper by Ahrland[21] and is presented in Table 10.5, which specifies the concentration of four major cations, nine minor metal ions and five anions. In order to calculate the complex species formed in this system it was necessary to employ the stability constants for all of the soluble complexes that may be formed as indicated in Table 10.6 as well as the solubility products of solids that may be formed. These values are listed in Table 10.7. The equilibrium constants listed in Tables 10.6 and 10.7 were obtained from Critical Stability Constants[1] and include estimated values where necessary to complete the data base. All of these constants were adjusted to seawater conditions (ionic strength 0.7 M) by the use of the correction factors given in Table 10.4 The results of these calculations may be expressed in various ways, but the computation provides

Table 10.5. Major and Some Minor Constituents of Seawater in Ahrland's Model[21]

Ion	Concentration M x 10^3	Trace Element	Concentration M x 10^9
Na^{2+}	479	Mn	4
Mg^{2+}	54.5	Fe	8
Ca^{2+}	10.5	Ni	5
K^{+}	10.4	Cu	4
Cl^{-}	559	Zn	5
SO_4^{2-}	28.9	Cd	0.1
HCO_3^-, CO_3^{2-}	2.35	Hg	0.02
Br^-	0.86	Pb	0.05
F^-	0.075	U	14

Table 10.6. Log Overall Stability Constants for Soluble Components of Seawater. μ = 0.70 M, t = 25.0°C.

$CaCO_3$	2.21	FeH^+_{-2}	-5.88	UO_2SO_4	1.80
$CaHCO_3^+$	9.90	FeH^-_{-4}	-20.76	$UO_2(SO_4)_2^{2-}$	2.50
$CaSO_4$	1.03	$Fe_2H^{4+}_{-2}$	-3.08	UO_2Cl^+	-0.10
CaF^+	0.60	$MnHCO_3^+$	10.00	UO_2F^+	4.4
CaH^+_{-1}	-12.20	$MnSO_4$	0.80	UO_2F_2	8.0
$MgCO_3$	2.05	$MnCl^+$	-0.20	$UO_2F_3^-$	10.55
$MgHCO_3^+$	9.80	$MnCl_2$	-0.3	$UO_2F_4^{2-}$	12.00
$MgSO_4$	0.90	$MnCl_3^-$	-0.50	UO_2Br^+	-0.3
MgF^+	1.30	MnH^+_{-1}	-10.80	$UO_2H^+_{-1}$	-5.8
MgH^+_{-1}	-11.70	MnF^+	0.70	$UO_2(SO_4)_3^{4-}$	3.7
KSO_4^-	0.30	$CuCO_3$	5.60	UO_2CO_3	8.9
$NaCO_3^-$	0.63	$Cu(CO_3)_2^{2-}$	7.90	$(UO_2)_3$-$(CO_3)H^+_{-3}$	-24.94
$NaSO_4^-$	0.33	$CuSO_4$	0.90	$CdCO_3$	2.00
$Fe(SO_4)_2^-$	2.00	$CuCl_2$	0.01	$CdCl^+$	1.35
$FeCl^{2+}$	0.64	CuF^+	0.80	$CdCl_2$	1.70

Table 10.6 continued

$FeCl_2^+$	0.70	$CuBr^+$	-0.20	$CdCl_3^-$	1.50
$FeCl_3$	-0.70	CuH_{-1}^+	-6.50	CdF^+	0.50
FeF^{2+}	5.20	$Cu_2H_{-2}^{2+}$	-14.80	$CdBr^+$	1.50
FeF_2^+	9.10	$(UO_2)_6H_{-12}(CO_3)_6^{12-}$	-10.46	$CdBr_2$	2.1
FeF_3^+	12.00	$(UO_2)_2H^{2-}$	-5.90	$CdBr_3^-$	2.5
$FeBr^{2+}$	-0.20	$UO_2(CO_3)_2^{2-}$	15.40	$CdBr_4^{2-}$	2.60
FeH_{-1}^{2+}	-2.76	$UO_2(CO_3)_3^{4-}$	20.3	CdH_{-1}^+	-9.90
CdH_{-2}	-20.0	$HgSO_4$	1.40	$PbCl^+$	0.90
CdH_{-3}^-	-31.0	$Hg(SO_4)_2^{2-}$	2.40	$PbCl_2$	1.30
$ZnSO_4$	0.90	$HgCl^+$	6.73	$PbCl_3^-$	1.4
$ZnCl^+$	0.01	$HgCl_2$	13.23	$PbBr^+$	1.08
$ZnCl_2$	-1.0	$HgCl_3^-$	14.20	$PbBr_2$	1.80
$ZnCl_3^-$	0.0	$HgCl_4^{2-}$	15.20	$PbBr_3^-$	2.10
ZnF^+	0.7	HgF^+	1.00	PbH_{-1}^+	-7.70
$ZnBr^+$	-0.40	$HgCO_3$	10.60	PbH_{-2}	-17.10
ZnH_{-1}^+	-10.20	$HgBr^+$	9.00	PbH_{-3}^-	-27.90
ZnH_{-2}	-19.40	$HgBr_2$	17.10	$Pb_2H_{-1}^{3+}$	-6.30
ZnH_{-3}^-	-27.70	$HgBr_3^-$	19.40	PbF^+	1.40
ZnH_{-4}^{2-}	-37.20	$HgBr_4^{2-}$	21.60	PbF_2	2.50
$ZnHCO_3^+$	10.50	HgH_{-1}^+	-3.70	HCO_3^-	9.54
$NiSO_4$	0.6	HgH_{-2}	-6.30	H_2CO_3	15.54
$NiCl^+$	-0.1	$HgHCO_3^+$	14.90	HSO_4^-	1.2
NiF^+	0.5	$PbCO_3$	6.20	HF	2.95
$NiBr^+$	-0.6	$Pb(CO_3)_2^{2-}$	8.80	H_{-1}^-	-13.76
NiH_{-1}^+	-10.10	$PbSO_4$	1.30		

Table 10.7. Log Solubility Products of Solids.
μ = 0.70 M; t = 25.0°C.

Phase	Log Product	Phase	Log Product
MnH_{-2}	14.78	PbF_2	-6.20
$Cu_2CO_3H_{-2}$	-6.00	PbH_{-2}	12.68
CuH_{-2}	9.10	UO_2CO_3	-12.90
$CdCO_3$	-9.60	UO_2H_{-2}	4.10
CdH_{-2}	13.48	$CaCO_3$	-6.60
$ZnCO_3$	-10.80	$CaSO_4$	-3.00
ZnH_{-2}	12.76	CaF_2	-10.60
$NiCO_3$	-7.00	CaH_{-2}	+22.41
NiH_{-2}	12.38	$MgCO_3$	-3.70
HgH_{-2}	1.58	MgF_2	-6.70
$PbCO_3$	-11.80	MgH_{-2}	16.43
$Pb_3(CO_3)_2H_{-2}$	-19.40	FeH_{-3}	2.77
$PbSO_4$	-6.0	$MnCO_3$	-8.70

a complete elucidation of the concentrations of free and complexed metal ions and ligands as a function of p[H].

If p[H] 8.1 is selected as an average or representative value for the conditions that apply in the open ocean the percent of metal ion associated with each anion may be calculated and is presented in Table 10.8. These data together with the data in Table 10.5 may be used to calculate directly the concentration of each complex species under the selected conditions. It is seen that the alkali metal ions show little tendency to complex the ligands available and the alkaline earth ions Ca^{2+} and Mg^{2+} are also primarily uncomplexed although about 10% of the Ca^{2+} ion is present as the sulfato complex while approximately 8% of Mg^{2+} ion is in that form. The high level of chloride ion in seawater is predominant in

Table 10.8. Seawater Speciation in the Absence of Added Ligands at pH 8.1 (Numbers give % in terms of Metal Ion Associated with the Ligands Indicated.)

Metal Ion	Cl^-	Br^-	F^-	SO_4^{2-}	$(H)CO_3$[a]	OH^-	Uncomplexed	Precipitate
Ca^{2+}			0.01	9.8	0.54		79.0	10.6 $CaCO_3$
Mg^{2+}			0.07	8.4	0.45	0.02	91.1	
K^+				2.2			97.8	
Na^+				2.4	0.01		97.6	
Mn^{2+}	34.4		0.01	4.4	0.15	0.12	60.9	
Cd^{2+}	96.8	0.08		0.3		0.05	2.9	
Hg^{2+}	99.9	0.05						
$Fe3^{+}$						1.71		98.3 $Fe(OH)_3$
Cu^{2+}	1.9			0.2	22.4	73.6	1.85	
Zn^{2+}	41.2	0.02		4.86	0.40	0.47	53.0	
Pb^{2+}	22.32	0.02		0.3	71.9	3.9	1.5	
Ni^{2+}	29.6	0.01		3.1		0.67	66.6	
UO_2^{2+}					100.0			

[a] $(H)CO_3$ carbonate and bicarbonate.

determining the speciation of Mn^{2+}, Cd^{2+}, and Hg^{2+} ions. Pb(II) is primarily present as a carbonato complex with smaller amounts of chloro and hydroxo complexes. A considerable fraction of the Ni(II) and Zn(II) ion are not complexed. The ferric ion seems to be almost completely converted to the colloidal hydroxide while the Cu(II) ion is present as a hydroxo complex in solution. There is considerable uncertainity about the speciation of the uranyl ion, UO_2^{2+}, because of the uncertainity in the equilibrium constants for the formation of possible complexes such as basic chlorides, which may compete with the carbonato complexes. On the basis of the data presently available it seems that all of the uranyl ion is converted to soluble uranyl carbonates.

With the determination of the speciation of metal complexes in the ocean as a baseline, it is now possible to determine how species distribution may be perturbed by EDTA. The stability constants for its complexes with the metal ions present in the ocean are indicated in Table 10.9. The values given were adjusted for ionic strength 0.7 M by taking the constants listed in critical stability constants[1] and applying the correction factors listed in Table 10.4. At an assumed concentration of 1.0×10^{-7} M of added EDTA the results of the calculation of the new distribution of complex species are given in Table 10.10. It is seen that at this concentration level EDTA has very little effect on the complexation of Ca^{2+}, Mg^{2+}, K^+, Na^+, Hg^{2+}, Fe^{3+} and UO_2^{2+}. On the other hand, because of their relatively high concentrations and the low concentration of added EDTA it is seen that 88% of the EDTA added is complexed by Ca^{2+} and Mg^{2+}. Redistribution of the complex concentrations present in natural seawater resulting from the addition of EDTA at 1.0×10^{-7} M level occurs only for those metal ions that form very stable EDTA chelates such as Ni(II) and Cu(II), which are almost completely converted to their EDTA chelates. The data available make possible calculations of the redistribution of metal complex species resulting from the addition of higher or lower levels of EDTA and this procedure may be extended to any other chelating agent for which the corresponding stability constants are known.

Table 10.9. Stability Constants of Metal Complexes of EDTA (expressed as log β values). μ = 0.70 M; t = 25.0°C.

	EDTA [a]			
	ML	MHL	$MH_{-1}L$	Other
Ca^{2+}	9.60	13.0		
Mg^{2+}	7.8	11.8		
K^{+}				
Na^{+}	1.2			
Fe^{3+}	23.8	25.1	17.0	
Mn^{2+}	12.8	13.0		
Cu^{2+}	17.7	20.8	5.8	
Cd^{2+}	15.4	18.4		
Zn^{2+}	15.4	18.5		
Ni^{2+}	17.5	20.8		
Hg^{2+}	20.5	23.7	10.8	
Pb^{2+}	16.9	19.8		
UO_2^{2-}	9.6	15.6		b

[a] Log protonation constants of EDTA (ethylenediaminetetraacetic acid) 8.78, 15.01, 17.39, 19.39. [b] log $\beta_{(UO_2)_2EDTA}$ = 16.5.

For a seawater sample with the composition indicated in Table 10.5 the number of components (including water) is 19. With the reactions that are indicated by the equilibrium constants in Tables 10.6 and 10.7 the number of soluble species is 141 while 26 insoluble compounds may also be present. This

Table 10.10. Seawater Speciation in the Presence of 1.0×10^{-7} M added EDTA at pH 8.1 (Numbers give % in terms of Metal Ion Associated with the Ligands Indicated

Metal Ion	Cl^-	Br^-	F^-	SO_4^{2-}	$(H)CO_3$ [a]	OH^-	Uncomplexed	Precipitate	EDTA
Ca^{2+}			0.01	9.8	0.54		79.0	10.6 $CaCO_3$	<0.01
Mg^{2+}			0.07	8.4	0.45	0.02	91.1		<0.01
K^+				2.2			97.8		<0.01
Na^+				2.4	0.01		97.6		<0.01
Mn^{2+}	34.1		0.01	4.4	0.15	0.12	60.3		0.9
Cd^{2+}	82.4	0.07		0.21		0.04	2.4		14.82
Hg^{2+}	99.9	0.05							<0.01
$Fe3^+$						1.71		98.2 $Fe(OH)_3$	0.1
Cu^{2+}	0.09				0.95	3.2	0.08		95.7
Zn^{2+}	9.8			1.2	0.1	0.1	12.6		76.3
Pb^{2+}	5.8			0.09	18.73	1.0	0.4		74.0
Ni^{2+}	0.06						0.13		99.8
UO_2^{2+}					100.0				<0.01

[a] $(H)CO_3$ carbonate and bicarbonate. [b] % of EDTA complexed: Ca^{2+} 79.7; Mg^{2+} 7.56; Mn^{2+} 0.04; Cd^{2+} 0.01; Cu^{2+} 3.83; Zn^{2+} 3.81; Pb^{2+} 0.04; Ni^{2+} 4.99.

treatment assumes that all possible complexes involving the components in Table 10.5 are formed at least to a slight extent. Very minor components such as complexes of Co^{2+} or non-complexing species such as borate and selenium were not considered. With an added chelating agent as a contaminant the number of components increases by 1, to 20, with a corresponding increase in the number of soluble species. In the case of EDTA the soluble species increases to 174 while the number of insoluble species considered remains the same at 26.

The species distributions presented in Tables 10.8 and 10.10 should be considered only a rudimentary model for speciation of metal complexes in the ocean. A complete model would include not only complexes in solution and equilibria with solid phases, but would also involve the kinetics of the processes together with the kinetics and equilibria of the various particulate adsorption and desorption phenomena. Such a model should also include the exposure to oxygen and the various redox processes possible in sea water under anaerobic conditions and in the presence of varying concentrations of oxygen. Also a complete model would take into account the complexities arising from the identification and inclusion of the wide variety of organic components that may be present.[22,23] These organic components include many complexing agents but are frequently of ill-defined nature. Considerable work needs to be done on metal ion affinities and reaction kinetics with particulate and organic systems before a more complete determination of metal speciation under various conditions encountered in seawater can be carried out.

This simple sea water model serves only to provide an example showing how the data base and speciation calculations may be used. These methods may be applied to the more complex natural environmental solutions containing many additional ligands and adsorbing particulate matter, provided that the appropriate equilibrium parameters are available, can be measured, or can be estimated with reasonable accuracy.

References

1. Smith, R. M.; Martell, A. E. "Critical Stability Constants" Vol. 1, 2, 3, 4, 5; Plenum: New York, 1974, 1975, 1977, 1976, 1982.
2. Sillen, L. G.; Martell, A. E. "Stability Constants, Special Publication No.17"; The Chemical Society: London, 1964.
3. Sillen, L. G.; Martell, A. E. "Stability Constants, Supplement No.1, Special Publication No.25"; The Chemical Society, London, 1971.
4. Harris, W. R. J. Coord. Chem., **1983**, 13, 16.
5. Larsson, E. Z. Phys. Chem., **1934**, A169, 208.
6. Bjerrum, J. Chem. Rev, **1950**, 46., 381.
7. Calvin, M.; Wilson, K. W. J. Am. Chem. Soc., **1945**, 67, 2003.
8. Calvin, M.; Melchior, N. C. J. Am. Chem. Soc., **1948**, 70, 3270.
9. Irving, H.; Williams, R.J.P. Nature (London), **1948**, 162, 746.
10. Irving, H.; Williams, R.J.P. J. Chem. Soc., **1953**, 3192.
11. Davies, C. W. J. Chem. Soc., **1951**, 1256.
12. Martell, A. E.; Calvin, M. "Chemistry of the Metal Chelate Compounds"; Prentice Hall: New York, 1952.
13. Van Uitert, L. G.; Fernelius, W. C.; Douglas, B. E. J. Am. Chem. Soc., **1953**, 75, 2736.
14. Irving, H.; Rossotti, H. Acta Chem. Scand., **1956**, 10, 72.
15. Nieboer, E.; McBryde, W.A.E. Can. J. Chem., **1970**, 48, 2549; 2565.
16. Smith, R. M.; Martell, A. E.; Motekaitis, R. J. Inorg. Chim. Acta, **1985**, 99, 207.
17. Smith, R. M.; Motekaitis, R. J.; Martell, A. E. Inorg. Chim. Acta, **1985**, 103, 73.
18. Smith, R. M.; Martell, A. E. Sci. of the Total Environ., **1987**, 64, 125.
19. Sillen, L. G. Science, **1967**, 156, 1189.
20. Motekaitis, R. J.; Martell, A. E. Marine Chem., **1987**, 21, 101.
21. Ahrland, S. "U.S.-Italy Workshop on Environmental Inorganic Chemistry", Eds. Irgolic, K. J.; Martell, A. E. VCH Publishers: Deerfield Park, Florida, 1985, pp.65-88.
22. Berman, S. S.; Yeats, P. A. CRC Critical Reviews in Anal. Chem., **1985**, 16, 1.

23. Martell, A. E.; Motekaitis, R. J. "Proceedings X International Symposium - Chemistry of the Mediterranean", Primosten, Yugoslavia, May 4-12, 1988.

APPENDIX I

PROGRAM PKAS

Program PKAS is a fast computer program for the calculation of protonation constants from potentiometric data. Its written if FORTRAN 77 and can be compiled as-is using either a personal computer or the VAX. It is an interactive program which takes the titration data from a file named FOR004.DAT and the original titration parameters interactively from the console or optionally from the file FOR003.DAT. It then immediately, refines the protonation constants, and readies output to be optionally written to disk on FOR001.DAT (which must be present since it is OPENED as an "OLD" file. At the discretion of the user, OPTIONS can be chosen which allow one to re-refine the data in many ways.

FILE USAGE

Program PKAS uses one required input file, one optional input file and one optional output file.

FOR004.DAT - input file (titration data)
FOR003.DAT - input file (optional file for parameters).
FOR001.dat - output file (optional)

The FOR004.DAT input file (required):

Line 1: Entire line available for title, identification (alphameric format)
Line 2: Volume Base, pH (2F8.4)
Line 3: " " " "
. . . . .
. . . . .
Line Z . . . .
(Program looks for end-of-file, do not leave a blank line.)

The FOR003.DAT parameter input file (optional):

	ITEM	FORMAT	DEFAULT
Line 1:	MILLIMOLES LIGAND	(F10.6)	none
Line 2:	NORMALITY OF BASE	(F10.6)	0.1000 M
Line 3:	INITIAL SOLUTION VOLUME mL	(F10.6)	50.00 mL
Line 4:	NUMBER OF DISSOCIATABLE PROTONS (ON NEUTRAL LIGAND)	(1I2)	none
Line 5:	NUMBER ADDITIONAL PROTONATIONS	(1I2)	none
Line 6:	FRACTION OF "A" VALUE TO BE EXCLUDED FROM COMPUTATION ON EACH END OF INTERVAL	(F10.6)	0.100
LINE 7:	MILLIMOLES EXCESS ACID ADDED	(F10.6)	0.000 mL
LINE 8:	pH CORRECTION	(F10.6)	0.000
LINE 9:	pK_w (ion product for water)	(F10.6)	13.78
LINE 10:	PERCENT CO2 IN BASE	(F10.6)	0.000

LINE 11:PROTONATION CONSTANT $(CO3)^{=}$ (F10.6) Log(K1)=10.00
LINE 12:PROTONATION CONSTANT $(HCO3)^{-}$ (F10.6) Log(K2)= 6.16
(end-of-file)

The FOR001.DAT output file (optional usage):

As soon as the initial computation is completed (several seconds), the results are written to the screen. A choice at this point can be made to write the result to DISK, to TERMINAL or go to OPTIONS. The disc file is opened as FOR001.DAT and can be saved for later printing.

NOTES:

1) The FOR003.DAT (optional) file is constructed in the same order as the program inquires for the corresponding data. It is for the convenience of the user so that he would be spared the requirement of keying-in the necessary parameters. The program PKAS is menu-driven and the first decision the user is given is the choice of using the input menus or the optional file FOR003.DAT depending on one of three answers to MILLIMOLES: a positive number will continue with the menus, a zero will read the input from FOR003.DAT and any negative input will additionally show the parameter order expected in FOR003.DAT.
2) OPTIONS: after the initial computation is complete, there is a number of OPTIONS provided if a recomputation is desired. Typing the two digit number activates the OPTION whereas typing RETURN recomputes the protonation constants. The following codes are in OPTIONS:
   "13" STOP: Elegant way to get out of the program.
   "14" LOG K CHANGE: Allows to individually alter a given protonation constant to see what this action does to the recalculation. Of theoretical interest for the curious
   "15" REFINEMENT: Allows for the automatic adjustment of millimoles of ligand. It may reach false minima if DISCARD interval is >0. This option should be tried several times to see if a lower SIGMA could be obtained. Use values in the range 0.0002-0.001 for the requested increment.
   "16" NUMBER OF POINTS: Controls the initial,final, and steps of the calculation. Only the titration points within the "A" intervals corresponding to the protonation constant should be left in for the final refinement. Thus an H3L ligand should not have any data points present outside $0<A<3$. If an additional protonation constant is thought to be present and is specified, then the permissible points are subject to

-1<A<3. "16" allows this selection without the physical removal of such points from the FOR004.DAT file. Nevertheless it is better to restrict the range to valid points initially.

"17" CONVERGENCE: The refinement of protonation constants stops when the rate of convergence is sufficiently slow. A smaller value for CONVERGENCE will add a few more iterations.

"18" TOP OF PROGRAM: Same effect as running the program initially. Everything is reset.

"19" ITERATION LIMIT: A value of 99 is a safety cut-off and has the effect of allowing the iterating process to terminate by the rate of convergence limit. Values <99 will necessarily iterate for the specified number of iterations. Thus a 98 will necessarily run for 98 iterations. This is fine on the VAX, but may be slower on a PC.

"20" MILLIMOLES: Normally the millimoles of ligand is a known quantity. If some doubt exists (hygroscopic substance, uncertain molecular weight,etc.) "20" allows changing the value of millimoles of ligand.

"21" NORMALITY OF BASE: Can enter another value to see what the effect of base strength is on the refinement.

"22" INITIAL VOLUME: Can enter another value to see what the effect of the initial volume is on the refinement.

"23" PCT CO2 IN BASE: All alkali solutions contain some carbon dioxide. By systematic variation, a suitable correction may be made using "23". This option is available for the curious, but it is scientifically more rigorous to make up fresh solutions.

"24" PH CORRECTION: If a constant correction factor is necessary on the values of pH in FOR004.DAT, OPTION "24" may be used to accomplish this adjustment.

"25" KW: If an incorrect -log Kw is used, it may be adjusted here.

"26" MILLIMOLES ACID: If the added acid is deemed incorrect, then "26" can be used to enter a better value.

"29" DISCARD FACTOR: Any discard value on "A" interval ends may be entered here. A "0" could also be used and is recommended during the refinement of millimoles. (Millimole refinement is usually not recommended using the sole justification of reducing the fitting parameters). A value of 0.1 is good when steep inflections are absent. Otherwise use 0.2. (Good work is indicated if small values such as 0.05

don't produce any sharp deviations near steep-slope inflections.

"30" CONTINUE: or RETURN allows to go back to the recalculation and refinement of the log protonation constants reflecting the OPTIONS just elected.

3) If FOR003.DAT is elected for input, note that the items are arranged so that only the first four are necessary to be specified: MILLIMOLES, BASE CONCENTRATION, VOLUME, AND NUMBER OF DISSOCIATIONS.
4) The parameter default values take effect when specific values are not entered as input. This is handy, because hitting RETURN is quicker than entering the default value.

DISKETTE

Four files are provided for use with PROGRAM PKAS:

PKAS.FOR	FORTRAN source code
PKAS.EXE	Compiled for running on a personal computer. If machine is incompatible, may have to recompile and relink the code.
TREN4.DAT	Sample titration data to be copied into FOR004.DAT before running PKAS.
TREN3.DAT	Parameter data for use in optional non interactive mode to be copied into FOR003.DAT before using.

```
      PROGRAM PKAS
      REAL NB,MMACID,MMLIG,LOGK,K,NUM,KW
      INTEGER BELL
      CHARACTER*4 HEAD
      REAL K1CO2,K2CO2,K1,K2
      DIMENSION TC(191),VB(191),PH(191),H(191),SIGINT(191),SI(191)
      DIMENSION CNA(191),TL(191),A(191),VOLUME(191)
      DIMENSION NPI(191), STORPH(191),LOGK(191),K(191),BETA(191)
      DIMENSION AVEDEV(191),TITLE(20),PHD(191),HDSQ(191)
      DIMENSION PSIGMA(191),SIGMA(191),DEV(191)
      DIMENSION SUMDEV(191),XH(191),XPH(191),TH(191),U(21)
      DATA CRIT/.0007/,BELL/007/,NN/0/
      DATA HEAD/'LOGK'/,TIGHT/.00001/
      DISC=0.1
C ***** OUTPUT FILE FOR001.DAT
      OPEN(1,FILE='FOR001.DAT',STATUS='OLD')
C ***** LUNI=INPUT FROM CONSOLE;   LUNO=OUTPUT TO TERMINAL SCREEN
      LUNI=0
      LUNO=0
      LUNI=5
      LUNO=6
8054  CONTINUE
      I1=0
      NRD=LUNI
      LUN=LUNO
      WRITE(LUNO,8072)BELL
8072  FORMAT(20X,'PKAS PROGRAM BY RAMUNAS J. MOTEKAITIS',/
     1 ,20X,'05/28/88 VERSION (fortran 77)',A4,//,
     2 ' file usage:',/
     3 ,' FOR004.DAT INPUT  FILE (required)',/
     4 ,' FOR003.DAT INPUT  FILE (optional)',/
     5 ,' FOR001.DAT OUTPUT FILE (required)',//)
      WRITE(LUNO,1111)
1111  FORMAT('  "0" Millimoles Redirects Input to FOR003.DAT',
     1  /' "-1" Millimoles also shows FOR003.DAT Parameter Order')
1112  FORMAT(' Millimoles Ligand',/' Normality of Base',/,
     1 ' Initial Volume',/,' "n" of HnL',/' "m" of H(n+m)Lm+',/,
     2 ' Discard on "a-value"',/,' Millimoles Excess Acid',/,
     3 ' pH Correction',/,' pKw',/,' % CO2 in Base',/,
     4 ' pK2 (CO2)',/,' pK1 (CO2)',/,' ["m" and beyond are optional]'/)
      SCALE =1.0
      I7=0
      MM=99
      NCODE=0
      MMACID=0
      PHCORR=0
      PKW=13.80
      PCTCO2=0
      K1CO2=10.0
      K2CO2=6.16
      WRITE(LUNO,8001)
8001  FORMAT(10X'MILLIMOLES LIGAND ?')
1     IF(NRD.EQ.3) OPEN(3,FILE='FOR003.DAT',STATUS='OLD')
      READ(NRD,8000) MMLIG
8000  FORMAT(1F10.6)
      IF(MMLIG.LT.0) WRITE(LUNO,1112)
      IF(MMLIG.LE.0) NRD=3
      IF(MMLIG.LE.0) GO TO 1
      WRITE(LUNO,8002)
8002  FORMAT(10X,'NORMALITY OF BASE ?')
      READ(NRD,8000) NB
      IF(NB.EQ.0) NB=.1
      WRITE(LUNO,8003)
8003  FORMAT(10X,'INITIAL VOLUME IN MILLILITERS ?')
      READ(NRD,8000) V0
      IF(V0.EQ.0) V0=50.
      WRITE(LUNO,8007)
8007  FORMAT(10X,'NUMBER OF DISSOCIABLE PROTONS ON NEUTRAL LIGAND ?')
      READ(NRD,7999) N
      WRITE(LUNO,8008)
7999  FORMAT(1I2)
8008  FORMAT(10X,'NUMBER OF ADDITIONAL PROTONATIONS TO BE '
     X  ,'CONSIDERED ?')
      READ(NRD,7999,END=7998) NN
      WRITE(LUNO,8014)
8014  FORMAT(10X,' FRACTION OF "A" VALUE TO BE EXCLUDED '/
     X  10X,' FROM COMPUTATION ON EACH END OF INTERVAL 0.1 ?')
      READ(NRD,8000,END=7998) DINP
```

```
      IF(DINP.LT.0) DISC=0.0
      IF(NRD.EQ.3) GO TO 13
      WRITE(LUNO,8015)
8015  FORMAT(10X,'TYPE "0" TO SKIP REMAINING OPTIONS'
     X  ,/,10X,'TYPE "1" TO CONSIDER THEM ?')
      READ(NRD,7999) IOPT
      IF(IOPT.EQ.0) GO TO 7998
      WRITE(LUNO,8004)
8004  FORMAT(10X,'MILLIMOLES OF EXCESS ACID ADDED ?')
13    READ(NRD,8000,END=7998) MMACID
      WRITE(LUNO,8010)
8010  FORMAT(10X,'PH CORRECTION ?')
      READ(NRD,8000,END=7998) PHCORR
      WRITE(LUNO,8011)
8011  FORMAT(10X,'PKW ?')
      READ(NRD,8000,END=7998) PKW
      IF(PKW.EQ.0) PKW=13.78
      WRITE(LUNO,8012)
8012  FORMAT(10X,'PER CENT CO2 IN BASE ?')
      READ(NRD,8000,END=7998) PCTCO2
      IF(PCTCO2.LE.0.0) GO TO 7998
      WRITE(LUNO,8013)
8013  FORMAT(10X,'ENTER THE STEPWISE PROTONATION CONSTANTS '/
     X 10X,'FOR CO2.  NORMALLY 10.00 AND 6.16 AT 25 DEG. ?')
      READ(NRD,8000,END=7998) K1CO2
      READ(NRD,8000,END=7998) K2CO2
7998  CONTINUE
      if(nrd.eq.3) CLOSE (UNIT=3)
      NRD=LUNI
      LUN=LUNO
      DISCL=DISC
      DISCU=1.-DISC
      NNN=N+NN
      NNNP1=NNN+1
      SNM1=N-1
      SNNN=NNN-1
8060  CONTINUE
      IF(PKW.LE.0) PKW=13.78
      KW=10.**(-PKW)
      IF(PHCORR.NE.0) GO TO 8071
      IF(NCODE.EQ.30.OR.NCODE.EQ.15) GO TO 2003
8071  OPEN(4,FILE='FOR004.DAT',STATUS='OLD')
      READ(4,32,END=4) (TITLE(II),II=1,20)
32    FORMAT(20A4)
      I=1
3     FORMAT(2F8.4)
2     READ(4,3,END=4)VB(I),PH(I)
      NCR=I
      PH(I)=PH(I)+PHCORR
      I=I+1
      GO TO 2
4     CONTINUE
      CLOSE (4)
      IF(I1.NE.0) GO TO 2004
      I1=1
      I2=NCR
      ISTEP=1
2004  SN=N
2003  I7=I7+1
      M=1
      DO 5 I=1,NCR
      H(I)=10.0**(-PH(I))
      XH(I)=H(I)
      VOLUME(I)=V0+VB(I)
      CNA(I)=NB*VB(I)/VOLUME(I)
      TL(I)=MMLIG/VOLUME(I)
      TC(I)=CNA(I)*PCTCO2/100.0
      TH(I)=SN*TL(I)
      A(I)=CNA(I)/TL(I)-MMACID/MMLIG
      CNA(I)=CNA(I)*(1.-PCTCO2/100.)
5     CONTINUE
C*******MIDPOINTS OF INTERVALS FOR INITIAL GUESSES
      DO 7 L=1,NNN
      NPI(L)=1
7     CONTINUE
      NPI(1)=0
      L=1
      AU=1-NN
```

```
      DO 6 I=1,NCR
      IF(L.GT.NNN) GO TO 6
      IF(A(I).LT.AU)GO TO 502
      GO TO 503
502   NPI(L)=NPI(L)+1
503   IF(A(I).GT.AU) GO TO 504
      GO TO 505
504   L=L+1
505   IF(A(I).GT.AU) GO TO 506
      GO TO 6
506   AU=AU+1.
6     CONTINUE
      IF(I7.GT.1) GO TO 8070
      I=0
      DO 8 L=1,NNN
      IF(NPI(L).GT.(NPI(L)/2*2)) GO TO 507
      GO TO 508
507   I=I+1
508   I=I+NPI(L)/2
      IF(I.EQ.0) GO TO 509
      GO TO 510
509   I=1
510   STORPH(L)=PH(I)
      I=I+NPI(L)/2
8     CONTINUE
      BNNN=0
      DO 9 L=1,NNN
      N10=N+NN-L+1
      LOGK(L)=STORPH(N10)
      BNNN=BNNN+LOGK(L)
9     CONTINUE
8070  CONTINUE
      IF (NCODE.EQ.15) GO TO 99
      WRITE(LUN,8072)
      WRITE(LUN,33)(TITLE(II),II=1,20)
33    FORMAT(1X,20A4/)
      IF(PCTCO2.LE.0.0)GO TO 1023
      WRITE(LUN,1021)PCTCO2,K1CO2,K2CO2
1021  FORMAT(' PCT CO2 IN BASE ='13X,F5.2,/,
     X                 LOG PROT CONST 1 =',5X,F6.2,/,
     X'      LOG PROT CONST 2 =',7X,F5.2)
1023  CONTINUE
      WRITE(LUN,8020) MMLIG
8020  FORMAT(' MILLIMOLES LIGAND              ',F8.5)
      WRITE(LUN,8021) NB
8021  FORMAT(' NORMALITY OF BASE              ',F8.5)
      WRITE(LUN,8022) V0
8022  FORMAT(' INITIAL VOLUME                 ',F8.2)
      WRITE(LUN,8023) NCR
8023  FORMAT(' NUMBER OF DATA POINTS          ',I8)
      WRITE(LUN,8024) PHCORR
8024  FORMAT(' PH CORRECTION                  ',F8.3)
      WRITE(LUN,8025) PKW
8025  FORMAT(' PKW                          ',F10.3)
      WRITE(LUN,8026) MMACID
8026  FORMAT(' ADDED ACID (MMOLES)            ',F8.5)
      WRITE(LUN,8029) DISC
8029  FORMAT(' DISC INCREMENT                 ',F8.4)
      WRITE(LUN,8030) TL(1)
8030  FORMAT(' LIGAND CONCENTRATION AT V0   ',1P,E10.4,/)
      WRITE(LUN,100) (HEAD,L,L=1,NNN)
100   FORMAT(3X,'M',2X,'SIGMA'2X,A4,I2,12(2X,A4,I2))
      WRITE(LUN,8117)
8117  FORMAT(/)
      K1=10.0**K1CO2
      K2=10.00**K2CO2
99    DO 10 L =1,NNN
10    K(L)=10.0**LOGK(L)
C******F, FP AND XH
      I=I1
18    CONTINUE
      J=0
      IF(XH(I).LE.0.OR.XH(I).GE.TH(I)) XH(I)=H(I)
      IA=A(I)
      SA=IA
      DA=ABS(A(I)-SA)
      IF(DA.LE.DISCL.OR.DA.GE.DISCU) GO TO 96
17    CONTINUE
```

```
      SN=N
      DERB=1.+KW/XH(I)**2
      XB=CNA(I)+XH(I)-KW/XH(I)
      XB=XB-MMACID/VOLUME(I)
      IF(PCTCO2.EQ.O) GO TO 171
      A1=1.+K1*XH(I)+K1*K2*XH(I)**2
      A2 = K1*XH(I) +2.*K1*K2*XH(I)**2
      DERA1 = K1+2.*K1*K2*XH(I)
      DERA2 = K1 + 4.*K1*K2*XH(I)
      DERB=DERB+TC(I)*(A1*DERA2-A2*DERA1)/A1**2
      XB=XB+TC(I)*A2/A1
171   F =( TH(I) -XB)/TL(I)
      FPRIME = -DERB/TL(I)
      NUM=0
      DENOM = 1.0
      DERNUM  = 0
      DERDEN = 0
      P=1.
      IF(BNNN-NNN*PH(I).GE.20.0) P=1.E-20
      IF(BNNN-NNN*PH(I).LE.-20.) P=1.E+20
      DENOM=P
      DO 12 L=1,NNN
      SL=L
11    XHK=K(L)*XH(I)
      P=P*XHK
      SLP=SL*P
      SLPH=SLP/XH(I)
      NUM = NUM + SLP
      DENOM = DENOM + P
      DERNUM = DERNUM + SL*SLPH
      DERDEN = DERDEN + SLPH
12    CONTINUE
      RNOD=NUM/DENOM
      F = F -RNOD
      FPRIME = FPRIME-DERNUM/DENOM+RNOD*(DERDEN/DENOM)
      XHNEW=XH(I)-F/FPRIME
      J=J+1
      IF(J.GT.100) GO TO 96
      SBA=ABS((XHNEW-XH(I))/XH(I))
      IF(SBA.LE.CRIT) GO TO 16
      XH(I)=XHNEW
      GO TO 17
16    CONTINUE
      IF(XHNEW.GT.0.0 ) GO TO 96
      IF(XH(I).LE.0.0)  XH(I)=H(I)
      GO TO 96
96    CONTINUE
      IF(J.GT.100) XH(I) =H(I)
      I=I+ISTEP
      IF(I.GT.I2) GO TO 19
      GO TO 18
19    CONTINUE
      ILIMIT =(I2-I1)/ISTEP+1
      SUMSQ=0
      DO 20 I=I1,I2,ISTEP
      HD= (XH(I)-H(I))/H(I)*100.0
      HDSQ(I)=HD*HD
      SUMSQ=HDSQ(I)+SUMSQ
20    CONTINUE
      S=ILIMIT
      SIGMA(M)=SQRT(SUMSQ/S)
97    CONTINUE
      WRITE(LUN,28) M,SIGMA(M),(LOGK(L),L=1,NNN)
28    FORMAT(1H+,I3,F7.2,13F8.4/(10X,13F8.4))
      IF(M.EQ.1) GO TO 82
      IF(SIGMA(M).LT.0.01) GO TO 98
      GO TO 21
98    WRITE(LUN,30) BELL
30    FORMAT(1H0,'END OF REFINEMENT',A4)
72    CONTINUE
      PSUMSQ=0
      DEND=0
      DO 433 I=I1,I2,ISTEP
      XPH(I)=-ALOG10(XH(I))
      PHD(I)=PH(I)-XPH(I)
2001  CONTINUE
      PHDSQ=PHD(I)**2
      PSUMSQ=PHDSQ+PSUMSQ
```

```
433     CONTINUE
        PSIGMA(M)=SQRT(PSUMSQ/S)
8062    CONTINUE
        IF(NCODE.EQ.15) GO TO 8311
        WRITE(LUN,228) M,PSIGMA(M),SIGMA(M)
228     FORMAT(I4,' ITERATIONS',F11.6,' OVERALL STD DEV IN LOG UNITS',
       X  F10.3,' SIGMA IN %H ')
        WRITE(LUN,227) (L,LOGK(L),SIGINT(L), L=1,NNN)
227     FORMAT(1X,4HLOGK,I2,3H = ,F8.4,'(',F8.4,')')
        DO 8161 L=1,NNN
        IF(LOGK(L).GT.PH(I2)) WRITE(LUN,8162) L
8162    FORMAT(' WARNING:  LOGK(',I2,') EXCEEDS HIGHEST PH VALUE')
        IF(LOGK(L).LT.PH(I1)) WRITE(LUN,8163) L
8163    FORMAT(' WARNING:  LOGK(',I2,') IS LESS THAN LOWEST PH VALUE')
8161    CONTINUE
        WRITE(LUN,31)
31      FORMAT(1H0,3X'I',8X,'VB',9X,'A',8X,'PH',2X,
       X  'CALCD PH',5X,'DIFF'/)
        WRITE(LUN,26) (I,VB(I),A(I),PH(I),XPH(I),PHD(I),I=I1,I2,ISTEP)
26      FORMAT(I5,5F10.3)
        WRITE(LUNO,8067) SIGMA(M)
8067    FORMAT (20X,'SIGMA=',F12.3)
80671   WRITE(LUNO,8063)
8063    FORMAT(/,1X,'5=TERMINAL 1=DISK 0=OPTIONS ?')
        READ(LUNI,7999) JREV
        LUN=JREV
        IF(JREV.EQ.1) GO TO 8070
        LUN=LUNO
        IF(JREV.EQ.5) GO TO 8070
        IF(JREV.NE.0) GO TO 80671
        PHCORR=0.0
        GO TO 8056
21      CONTINUE
        IF(M.EQ.1.OR.MM.LT.99) GO TO 82
        IF(TIGHT.LT.0.00001.OR.TIGHT.GT.0.02) TIGHT=0.005
        TIGP1=TIGHT+1
        TIGP=TIGHT*100
        IF(NCODE.EQ.14) GO TO 72
        IF((ABS(SIGMA(M)-SIGMA(M-1))/SIGMA(M)).LT.TIGHT) GO TO 522
        GO TO 523
522     WRITE(LUN,81) TIGP
523     IF(SIGMA(M-1).LT.TIGP1*SIGMA(M)) GO TO 72
        GO TO 82
81      FORMAT(1H0,'  REFINEMENT TERMINATED BY CONVERGENCE LIMIT.'/
       X  ' SUCCESSIVE STANDARD DEVIATIONS ARE LESS THAN ',F5.3,'PER',
       X  'CENT APART'/)
82      CONTINUE
        M=M+1
        IF(M.GT.MM)GO TO 524
        GO TO 525
524     WRITE(LUNO,80) MM
80      FORMAT(1H0, 'REFINEMENT TERMINATED AT 'I3,'PASSES')
        SIGMA(M)=SIGMA(M-1)
525     IF(M.GT.MM) GO TO 72
526     DO 23 L=1,NNN
        SI(L)=0
        NPI(L) =1
        SUMDEV(L)=1.0
23      CONTINUE
        NPI(1) = 0
        AU=1 - NN
        L=1
        DO 22 I=I1,I2,ISTEP
        IA=A(I)
        SA = IA
        DA=ABS(A(I)-SA)
        IF(DA.LE.DISCL.OR.DA.GE.DISCU) GO TO 22
        IF(L.GT.NNN) GO TO 528
        GO TO 529
528     L=L-1
529     DEV(L)=H(I)/XH(I)
        SI(L)=SI(L)+ALOG10(DEV(L))**2
        WGHT= 1.-ABS(AU-A(I)-.5)/.5

        if(wght.lt.0)wght=-wght

        DEV(L)=DEV(L)**WGHT
        SUMDEV(L) = SUMDEV(L)*DEV(L)
```

```
      IF(A(I).LT.AU)GO TO 530
      GO TO 531
530   NPI(L)=NPI(L) +1
531   IF(A(I).GT.AU) GO TO 532
      GO TO 533
532   L= L+1
533   IF(A(I).GT.AU) GO TO 534
      GO TO 22
534   AU=AU+ 1.
22    CONTINUE
C*******COMPUTE AVERAGE DEVIATION FOR EACH INTERVAL
      DO 24 L=1,NNN
      SNPI= NPI(L)
      SNIP=SNPI-1
      IF(SNIP.LE.0) SNIP=1
      AVEDEV(L) = SUMDEV(L)**(1./SNPI)
      I10=NNNP1-L
      SIGINT(I10)=SQRT(SI(L)/SNIP)
24    CONTINUE
C*******ADJUST K'S ACCORDING TO AVERAGE DEVIATION
      BNNN=0
      DO 25 L=1,NNN
      N10=N+NN-L+1
      LOGK(L) = LOGK(L)-ALOG10(AVEDEV(N10)*SCALE)
      BNNN=BNNN+LOGK(L)
25    CONTINUE
      GO TO 99
8056  WRITE(LUNO,8057)I1,I2,ISTEP,TIGHT,MM,MMLIG,NB
8057  FORMAT(1H1,/,
     X  ' TYPE "13" FOR FORTRAN STOP',/,
     X  ' TYPE "14" FOR MANUAL LOGK CHANGE',/,
     X  ' TYPE "15" FOR REFINEMENT',/,
     X  ' TYPE "16" FOR NUMBER OF POINTS (',3I3,')',/,
     X  ' TYPE "17" FOR CONVERGENCE (',F7.5,')',/,
     X  ' TYPE "18" FOR TOP OF PROGRAM ',/,
     X  ' TYPE "19" FOR NEW ITERATION LIMIT (',I3,') ',/,
     X  '           (if N<99 then N will be done)'/,
     X  ' TYPE "20" FOR NEW MILLIMOLES (',F8.5,')',/,
     X  ' TYPE "21" FOR NEW NORMALITY BASE (',F8.5,') ')
      WRITE(LUNO,9057)V0,PCTCO2,PHCORR,PKW,MMACID,DISC
9057  FORMAT(' TYPE "22" FOR NEW INITIAL VOLUME (',F6.2,')  ',/,
     X  ' TYPE "23" FOR NEW PCT CO2 IN BASE (',F5.2,') ',/,
     X  ' TYPE "24" FOR NEW PH CORRECTION (',F6.3,')    ',/,
     X  ' TYPE "25" FOR NEW KW (',F7.3,')               ',/,
     X  ' TYPE "26" FOR NEW MMOLES ACID (',F9.5,')      ',/,
     X  ' TYPE "29" FOR NEW DISCARD FACTOR (',F7.4,')   ',/,
     X  ' TYPE "30" TO CONTINUE',/,
     X  '???')
      LUN=LUNO
      READ(LUNI,7997) SNCODE
7997  FORMAT(F2.0)
      NCODE=SNCODE
      IF(NCODE.EQ.0) NCODE=30
      IF(NCODE.EQ.13) STOP
      IF(NCODE.LT.14.OR.NCODE.GT.30) GO TO 8056
      IF(NCODE.EQ.27.OR.NCODE.EQ.28) GO TO 8056
      IF(NCODE.EQ.18) GO TO 8054
      IF(NCODE.EQ.30) GO TO 7998
      IF(NCODE.EQ.16) GO TO 8061
      IF(NCODE.EQ.15) GO TO 8299
      IF(NCODE.EQ.14) GO TO 8901
8059  WRITE(LUNO,8058)
8058  FORMAT(1X,' NOW TYPE IN NEW VALUE(F10.6) ?')
      READ(LUNI,8000) XXX
      IF(NCODE.EQ.17) TIGHT=XXX
      IF(NCODE.EQ.19) MM=XXX
      IF(NCODE.EQ.20) MMLIG = XXX
      IF(NCODE.EQ.21) NB    = XXX
      IF(NCODE.EQ.22) V0    = XXX
      IF(NCODE.EQ.23) PCTCO2 = XXX
      IF(NCODE.EQ.24) PHCORR = XXX
      IF(NCODE.EQ.25) PKW=XXX
      IF(NCODE.EQ.26) MMACID= XXX
      IF(NCODE.EQ.29) DISC  = XXX
      NCODE=10
      GO TO 8056
8061  WRITE(LUNO,8262) BELL
8262  FORMAT(1X,A2,'IIN,IFINAL,ISTEP(I2,I3,I2)?')
```

```
      READ(LUNI,8263) I1,I2,ISTEP
8263  FORMAT(I2,I3,I2)
      IF(I1.EQ.0) I1=1
      IF(I2.EQ.0.OR.I2.GT.NCR) I2=NCR
      IF(ISTEP.EQ.0) ISTEP=1
      ILIMIT=(I2-I1)/ISTEP+1
      GO TO 8056
8299  CONTINUE
      DF=1
      IR=1
      NSTOP=0
      U(IR)=SIGMA(M)
      SIGMIN=SIGMA(M)
      WRITE(LUNO,8301)
8301  FORMAT(' TYPE IN (a small!) MMOLE INCREMENT?')
8302  READ(LUNI,8000)DLB
      IF (DLB.EQ.0) GO TO 8302
8310  MMLIG=MMLIG+DLB*DF
      GO TO 7998
8311  IR=IR+1
      IF(IR.GT.20) GO TO 8056
      U(IR)=SIGMA(M)
      IF(NSTOP.EQ.1) GO TO 8400
8190  CONTINUE
      IF(NSTOP.EQ.-3) GO TO 8450
      IF(IR.EQ.2) GO TO 8250
      IF(NSTOP.EQ.-2) GO TO 8260
      IF(IR.LT.3) GO TO 8310
8195  CONTINUE
      IF(U(IR).LT.U(IR-1).AND.U(IR-1).LT.U(IR-2)) GO TO 8310
      IF(U(IR).GT.U(IR-1).AND.U(IR-1).LT.U(IR-2)) GO TO 8200
      IF(U(IR).GT.U(IR-1).AND.U(IR-1).GT.U(IR-2)) GO TO 8300
      GO TO 8310
8200  X=-.5*(U(IR-2)-4.*U(IR-1)+3.*U(IR))/(U(IR-2)-2.*U(IR-1)+U(IR))
      WRITE(LUNO,8201) ir,x
8201  FORMAT(//,'ir  x ',i5,f16.6//)
      DF=DF*X
      NSTOP=1
      IR=0
      GO TO 8310
8250  IF(U(1).GT.U(2)) GO TO 8310
      HOLD=U(2)
      U(2)=U(1)
      U(1)=HOLD
      DF=-2.*DF
      NSTOP=-2
      GO TO 8310
8260  NSTOP=0
      DF=DF/2
      GO TO 8195
8300  HOLD=U(IR)
      U(IR)=U(IR-2)
      U(IR-2)=HOLD
      DF=-3.*DF
      NSTOP=-3
      GO TO 8310
8400  DF=DF/X
      DF=.1*DF
      IF(ABS(DLB*DF).LE.(0.0002*MMLIG)) GO TO 8500
      NSTOP=0
      GO TO 8190
8450  NSTOP=0
      DF=DF/3
      GO TO 8195
8500  CONTINUE
      NAUTO=1
      WRITE(LUNO,8467) BELL
8467  FORMAT(1X,A4,'END OF REFINEMENT'/)
      NCODE=30
      GO TO 7998
8901  WRITE(LUNO,8902) (L,LOGK(L),L=1,NNN)
8902  FORMAT(' LOGK(',I2,')=',F9.4)
      WRITE(LUNO,8903) BELL
8903  FORMAT(1X,A4,'ENTER I=?')
8904  READ(LUNI,7999) L
      IF(L.GT.NNN) GO TO 8904
      IF(L.LT.1) GO TO 99
8906  WRITE(LUNO,8905) L,BELL
```

```
8905  FORMAT(1X,'LOGK(',I2,')=?',A4)
      IF(LOGK(L).EQ.0) GO TO 8906
      READ(LUNI,8000) LOGK(L)
      GO TO 8901
      END

C      LAST LINE OF PROGRAM PKAS.FOR
```

APPENDIX II

PROGRAM BEST

Program BEST is a FORTRAN computer program for the computation of titration curves from user-supplied input data. Because it is interactive, it is a very powerful tool for the computation of stability constants of practically any degree of complexity. In this regard, BEST is used to minimize the standard deviation of fit (SIGFIT) between observed and calculated pH values of the entire titration curve. This minimization is accomplished through either manual or automatic variation of selected stability constants. In addition, fine tuning is possible through the minimization in SIGFIT by the variation of any of the input parameters which define the mathematical shape of the titration curve. SIGFIT is computed using numerically calculated square reciprocal slope as a weighting factor so as to increase the sensitivity of the computations to buffer regions and to diminish the importance of potential breaks (inflections).

<u>FILE USAGE</u>

Program BEST uses one input file, one output file, and 3 logging (scratch) files:

- FOR004.DAT - input file (remains unchanged)
- FOR003.DAT - scratch file (stores current log Betas)
- FOR002.DAT - scratch file (stores current millimoles)
- FOR001.DAT - output file (results for hard copy)
- ERROR.DAT - scratch file (stores session errors in log Betas)

<u>INPUT FILE</u>

The name of the input file which BEST recognizes FOR004.DAT. It has a five block structure. Two of the variable blocks are ended by the presence of a blank line. Fixed blocks are 1 line in length. The fifth block does not end with a blank (BEST knows when the end-of-file comes).The FOR004.DAT file is constructed in general:

- BLOCK 1: one line of 80 characters of free format (80A1)
- BLOCK 2: one line for each component name and quantity in millimoles - NAME(J) MMC(J) (8A1,1F8.5)
- BLANK LINE: interpreted as end-of-component list
- BLOCK 3: one line for initial volume (mL), concentration of base, quantity in millimoles of added strong acid, and pH correction increment to be applied to all readings - VO NB MMACID PHCOR (4F8.5)
- BLOCK 4: one line for each species log β, component number and stoichiometry - LBETA(K) IC(K) IS(I,K) (F8.5,14(2I2,1X))
- BLANK LINE: interpreted as end-of-species list

BLOCK 5: one line for each titration point in terms of volume base in mL and pH [and list of -log(component concentrations) - VB(I) PH(I) RCC(I,K) (10F8.4)

EOF: Do NOT leave any blank lines here !

Symbolically recapitulating the FOR004.DAT input file layout:

```
BLOCK 1 Line 1: Title line (80A1)
BLOCK 2 Line 2: Component #1, Millimoles (8A1,F8.5)
        Line 3: Component #2,  "     "     "    "
        .       .       .       .       .       .
        .       .       .       .       .       .
        Line x: Blank line to flag end of component list
BLOCK 3 Line x+1: V0, NB, MMACID, PHCOR (4F8.5)
BLOCK 4 Line x+2: Log(B1),C1S1,C2S2,.....,CHSH (F8.5,H(2I2,1X))
        Line x+3: Log(B2),C1S2,C2S2,.....,CHSH (F8.5,H(2I2,1X))
        .       .       .       .       .      (F8.5,H(2I2,1X))
        .       .       .       .       .      (F8.5,H(2I2,1X))
        Line y: Blank line to indicate end of species list
BLOCK 5 Line y+1: mL base,pH,Log([1]),Log([2]),...(2F8.5,HF8.5)
        Line y+2: mL base, pH (2F8.5)
        Line y+3: mL base, pH (2F8.5)
        .       .       .       .
        .       .       .       .
        Line z: last mL base, pH (2F8.5)
        Do NOT leave any blank lines here !
```

Definitions:

x: line number depending on H the number of components
y: line number depending on H the number of components and number of species
z: final line number containing the last data point
V0: initial volume in mL of solution
NB: normality of strong base used
MMACID: millimoles of excess acid present but not included in final component specification
PHCOR: added value to calculated p[H]
B1: overall stability constant of species 1
B2: overall stability constant of species 2
C1: component number, i.e. 01
C2: component number, i.e. 02
S1: stoichiometric coefficient of C1 in species
S2: stoichiometric coefficient of C2 in species
H: number of chemical components
[1]: log concentration estimate of deprotonated component 1
[2]: log concentration estimate of deprotonated component 2
EOF: end-of-file

Examples of lines:

```
Line 1  :Page 49, Second Try, Dried Edta
Line 2  :EDTAbbbb.1733
Line 3  :Ca++bbbb.1722
Line 4  :Protbbbb.3466
Line x  :bbbbbbbbbbbbbbbbbbbbbbbbbbbbbbbbbbbbbbbbbbbbbbb
Line x+1:50.00bbb.1011bbb.2001bbb-.003
Line x+2:0bbbbbb0101b0200b0300
Line x+3:12.99   0101 0201 0300
Line x+4:6.2     0101 0201 03-1
Line x+5:-13.78  0100 0200 03-1
Line y+1:0.0bbbbb2.301bbb-12.0bbb-4.0
Line y+2:0.1     2.378
Line y+3:0.2     2.436
Line z  :9.0     10.456
```

Notes:

b's are symbolic for the presence of blank spaces. Do not type a 'b'.

Avoid using TAB. Always use spaces.

The component HYDROGEN is always the last component.

.1733 millimoles of EDTA.

.1722 millimoles Ca^{++}.

.3466 millimoles H+ associated with ligand (=2x.1733).

"EDTA" " Ca++" "Prot" are arbitrary component labels.

50.00 mL initial volume before any base was added.

.1011 molar strong base titrant.

.2001 millimoles of strong acid added prior to titration

-.003 pH correction based on meter calibration (this pH meter reads 0.003 units low relative to what it should).

12.99 log of normal stability constant, species ML.

6.2 log of overall stability constant, species $MH_{-1}L$.

0 species L in its most deprotonated state.

-13.78 log of water ion product, species OH^-.

-12.0 log of $EDTA^{4-}$ concentration guess.

-4.0 log of metal ion concentration guess

0.0 mL base

2.301 pH

0.1 mL base

2.378 pH

0.2 2.436 : next data point

9.0 10.456: last line, last data point

<u>OUTPUT FILE</u>

The principal output file from BEST is called FOR001.DAT and will contain any results deliberately placed into it by the user. It could contain stability constants, a-values, results, species concentrations, or combinations of choices. One common use of FOR001.DAT is for obtaining a hard copy of the refinement.

SCRATCH FILES

Best uses FOR002.DAT and FOR003.DAT for storing current millimole values and log Beta values respectively during the refinement of the constants. These values are updated at each completed refinement cycle and can be called up out of OPTIONS. They are extremely useful in case of program crashes or after pausing between program sessions. FOR003.DAT is useful in passing the latest log Beta's when working with several sets of data. Since they are updated during refinement, FOR002.DAT and FOR003.DAT must be initially present on disk (even if blank, such as when starting fresh).

ERROR.DAT FILE

This is a logging file and is generated and used as a log file during ERROR analysis. It will be normally blank unless the user attempts to use the ERROR option.

DIRECTIONS

1: Construct input file FOR004.DAT
2: Construct blank scratch files FOR002.DAT and FOR003.DAT (used to store current millimoles and log Betas).
3: RUN BEST
4: ECHO: See line by line FOR004.DAT during initial input. Useful for pinpointing the line containing typos in the input file. Responses: Y or RETURN
5: OPTIONS Menu: Recommended response first time: RETURN (or 24 or 63). This sets into motion a first pass calculation.
6: Main Menu: "27=TOP 36=OUTPUT 45=REFINE 54=OPTIONS 63=COMPUTE 72=STOP". RETURN can be used to toggle between the OPTIONS Menu and Main Menu.
   27 TOP: Same effect as RUN BEST
   36 OUTPUT: Enables various output combinations: "5=TERMINAL 1=DISK 0=RESULTS"
      1=DATA: Title, Parameters, Overall Constants
      2=a VALUES: Moles base/mole of each component
      3=SPECIES: Species distribution by species or by percentage for selected data points
      4=RESULTS: Computed titration curve and SIGFIT
      Recommended response: 5 (=TERMINAL) to preview what combinations of output are desired for writing to FOR001.DAT. The response 1 (=DISK) writes all further requests to FOR001.DAT for later printing.

45 REFINE: Enables refinement option for log Beta values: (Do a 63 before invoking 45 the first time) "-CYCLES;ITEMS/CYCLE(99=repeat menu);[1]=(ALL);[ITER LIMIT]"

Response examples to "45" with explanations:

0302 refine two constants in succession a total of 3 times (3 cycles).

030201 same but additionally display the progress of each refinement step. (This display is nice but is more CPU-time intensive.)

03020110 same but additionally reset the default 5 iterations/cycle to 10. On slow computers leave at 5 otherwise 10 is a good compromise. Max=17 allowed.

019901-1 1 cycle, previous menu, display progress, but change only the increment. (saves typing menu)

0599 5 cycles with previous menu.

"-ENTER I of BETA AND COVARYING I's" response examples:

05 The fifth overall constant chosen for refinement

050607 The fifth, sixth, and seventh constants will be refined together while keeping the ratios of their stepwise representations fixed. (There will be as many responses without further prompt as the number specified in the ITEMS/CYCLE above. Recommended response: A single constant representation per line for simple systems and covarying combinations for complex, stepwise equilibria if single entries prove unsatisfactory.)

"-ENTER INCREMENT: Suggestions: .1 is usually fine at the beginning. 1.0 may be advisable in case the magnitude of the log Beta is not known. Smaller increments should be reserved for cases where refinement is essentially complete, otherwise computer will run for a long time.

54 OPTIONS: Enables manual or automatic alteration of titration parameters and manual changes of specific log Betas.

'15' restores beep prompt

'16' SILENT MODE (turns off beep prompt)

'17' SPEED FACTOR: Acceptable values: -1,0-5 (In 'troublesome situations' a -1 will here reveals the deprotonated concentration and mass-balance residual while solving each titration point.)

0: normal, best left alone.

1: provides faster running by ignoring the minor species.

5: provides ca. 30 % speed at the expense of some accuracy. Sometimes blows up even at 3.

'18' IIN,IFINAL,ISTEP: Used to redefine the range and steps of the computation. Default is 1,N,1, where N is the number of titration points.

Response examples to '18' with explanations:

01-101 Consider N (all) points sequentially

013003 Consider 1,4,7,10,13,16,19,22, 25,28 (speeds rough refinements by skipping points)

30-102 Starting at 30 consider every other point until N is reached.

000001 Keep initial and final, just change step

'19' Alter BETA values:

1) Enter species number I
2) Enter new value for log Beta(I)
3) RETURN or 0 to get back to OPTIONS

'20' Alter Component MILLIMOLES

1) Enter component number I
2) Enter new value for millimoles of component I
3) RETURN or 0 to get back to OPTIONS

'21' Alter NORMALITY of BASE "Now type in Value"

Response: new number

'22' MILLIMOLES EXCESS ACID (in addition to that in '20') "Now type in value" Response: new value (Total titratable acid is the sum of this value plus the amount shown in (20) MILLIMOLES list. Do NOT duplicate here what has been declared in (20).

'23' INITIAL VOLUME (mL in titration vessel at start of titration) "Now type in value" Response: new value

'24' (or '63' or RETURN) to exit OPTIONS 54

'25' PARABOLIC FUNCTION A Convenience utility for making an improved judgement of the minimum based on 3 pairs of parameters,(say millimoles (1) and sigma (2) in this example:

param 1	param 2
.10000	.028456
.10010	.018765
.10020	.019009

Interpolated minimum predicted .100145

Predicted sigma .018264 (Useful when parabolic minimum is desired)

'26' APPORTION HYDROGEN Realigns millimoles hydrogen as an integral multiple of ligand while depositing the remainder into excess acid. This alignment allows to make sense of the a-values. Eg. Suppose Ga(III) and EDTA (H_4L) are listed as components 1 and 2. The suggested response would be 0204 for the component number and multiplier.

'27' Each refinement ('45') updates Beta's on disk. This option reads these values form FOR003.DAT into program memory.

'28' Each refinement ('45') updates Millimoles on disk. This option reads the FOR002.DAT file into memory.

'29' PH CORRECTION: A constant to be added to each calculated pH. When the meter reads high, enter positive difference; when meter reads low enter a negative difference. "Now type in value" Response: new value

'30' AUTO-REFINE MILLIMOLES: The millimoles of selected components can be automatically adjusted for BEST fit to the experimental data with or without simultaneous refinement of log Beta's. DO NOT USE without good reason or proper justification.

1) To be used only when one or more millimole quantities are unknown. Adjustment of the millimoles of hydrogen may be justified when the exact zero point of titrant is in doubt or its exact amount present is in doubt such as when cumulative error may be present. Egs: Hygroscopic ligand, excess acid present in metal stock solution, and excess acid added to the experimental solution.
2) Increment should be small. (i.e. 0.0001-0.0020)
3) If more than one component is refined run times can be many hours.
4) Use prudence: limit to only 1 or 2 (not recommended) cycles on Betas.
5) CAUTION: Do not deposit experimental errors into millimole adjustment.
6) When prompted for CYCLES it is ok to respond 0199 if a protocol had already been established in REFINE 45
7) It is also proper to hit RETURN in order to limit refinement to millimoles alone and thus avoid the simultaneous log(Beta) refinement.

'31' ERROR ANALYSIS Computes the error estimate on selected Log Betas based on the incremental effect on SIGMA of incrementally changing that Beta value. (It is most rigorously meaningful when the basis for this computation is limited to the pH interval where the species exists to say > 5 %. When the entire pH profile is used, the error tends to become overestimated.) A disk file ERROR.DAT will be generated with an entry for each attempt but only the last results for each log Beta will be listed using OUTPUT (36).

63 COMPUTE: Computes a new titration curve with the current parameters and prints SIGFIT. After changing parameters, a 63 should be done before REFINE (45). (Sometimes 63 should be issued several times in succession to ensure the calculation of each point of the titration. This difficulty arises in cases of exact 1.0000:1.0000 stated stoichiometry while formation is >99.999 % complete and is related to single precision floating point representation.)

72 STOP: Exits BEST gracefully closing all files.

REFINE (45) STRATEGY:

(Usually it is BEST to compute all refinements using the entire titration curve. This avoids the common pitfall of refining a portion of the curve while accumulating deviations in unseen portions of the curve. On slow computers one can do every third or fifth data point). (It is of upmost importance that there be a finite concentration of each equilibrating species when refining its respective constant. Use the SPECIES TERMINAL output mode to check for finite concentration).

Simple case: When only one log BETA value is unknown, simply enter REFINE mode, 0101 (or 010101 to view progress), enter the species number (say 05) (or numbers if it is desired to co-vary other values (say 050607), enter an increment (say .4). Here's what happens. The curve will be computed with a value of log BETA + 0.4000. If the SIGFIT is greater than before, then the curve will be recomputed with a value of log BETA - 0.4000. This process of incrementation and computation will continue for up to 5 times until a minimum SIGFIT is found. A parabolic interpolation of the log BETA vs SIGFIT is made and the minimum is located. The titration curve is recomputed and SIGFIT(1) is computed. (On the other hand, if a minimum SIGFIT is not found in the 5 tries, SIGFIT(6) is computed and the computation stops. At this point, usually it is most convenient to re-enter REFINE mode with a 0199 (or 019901).) Because the increment (0.4) is very large, enter a 019900-1 (or 019901-1) to change it to 0.01 or so. If SIGFIT(1) is computed then the refinement has been completed. If SIGFIT(6) is the final computation then do 0199.

An alternate approach to this example is to enter 01010117 followed by the species number(s) with the increment 0.100 thus allowing many small steps instead of several larger steps. The instruction 01010117 changes the default limit to

17 the number of steps of 0.1 BEST tries before issuing another choice. On a slower computer this haphazard commitment to many computations is very inefficient. It is also inefficient to, especially in simple cases to enter as a first instruction say 100101, because BEST will step through incrementations on the log Beta necessarily for 10 cycles, even if it had found the minimum on the first cycle.

Complex case: When two or more constants are unknown, do not start as in the simple case, one constant at a time. (Assume log Beta for species 07, 08, and 13 are unknown.) Now enter REFINE, 020301,07,08,13, and .1 in succession. (020301 means two complete refinement cycles with three log BETAS minimized in each refinement cycle while viewing the progress of the process.) ".1" represents a suggested modest starting increment, but depending on the degree of convergence could have been smaller. If SIGFIT is progressively decreasing but does not reach minimum in two cycles, reenter REFINE with say 0599, for say five cycles with the same menu. (Or 059900-1, if you want to change the increment.) In order to save time, it is BEST not to request too many cycles, before the rate of progress is ascertained. Inspect the rate of convergence. If the refinement rate (as measured by successive SIGMA(1) differences) is becoming slow, decrease the increment for smoother results. There are no arbitrary cut-offs built in. Many overlapping constants require many more cycles than several-constant problems. Therefore it is left up to the investigator to judge the completeness of the refinement. In practice, a point is always reached when the successive SIGFIT values stay constant and the stability constants at successive cycles remain fixed. Further refinement is then not possible. (In general it is very inefficient to enter REFINE with a menu of log BETAS and to simply set a large number of ITERATIONS. The danger lies in that the first log BETA considered will attempt to artificially minimize the SIGFIT in an extreme way compensating for the other unknown log BETAS in such a way as to either cause the first log BETA to become insignificantly small, or chemically non-sensibly large, such that when the remaining log BETAS are considered, their values may become also artificially distorted so that ultimate refinement may even become impossible. For this reason you may want to keep the iterations per cycle at 5, at least initially.)

Parameter refinement: While good analytical chemistry practice has it that the values of all titration parameters be known with good precision, occasions arise when one or more of these parameters are NOT known with as much accuracy as one would like. Some examples:

-Gallium is prepared by the dissolution of the weighed metal into 12 M Hydrochloric acid. Most of the excess acid is gently evaporated off and after quantitative dilution with water the Gallium(III) concentration is checked with standard EDTA. Although the residual free acid may be estimated via graphical or titrimetric methods, its value may be obtained using OPTIONS '22' (excess acid) by trial and error coupled with '26' (parabolic interpolation) or using OPTIONS '30' automatically.
-A polyamine polyhydrobromide sample recrystallized from an alcoholic HBr solution. The expected polyhydrobromide would consist of crystals containing up to 3% of occluded excess HBr whose amount could be assessed within OPTIONS.
-A ligand synthesized as an organically pure (NMR, MS) oil containing some salt, water and an indefinite ratio of mineral acid. Here the millimoles of both ligand and acid could be co-refined while determining the protonation constants of the ligand.

CAUTION: IT CANNOT BE OVEREMPHASIZED THAT THE REFINEMENT OF ANALYTICAL PARAMETERS SHOULD BE JUSTIFIED IN EACH CASE RATHER THAN BECOME ROUTINE PRACTICE SIMPLY TO MAKE SIGMA FIT LOOK BETTER.

HINTS

1. Initial guesses of components are important. The required number of guesses is one less than the number of stated components. Each guess is the log value of the concentration of the deprotonated form of the component. Because titrations usually start in the acid region, typical numbers are -3.0 or -4.0 for metal ions and -15.0 to -30.0 for ligands. These guesses are enumerated at the first titration point in fields of 8 characters. For back- titrations, the metal is usually totally complexed at high pH and any free ligand is in a deprotonated form, thus their respective initial guesses would be near -10.0 to -20.0 and -4.0 to -8.0 or so. In case of bad starts, these values must be varied in some sort of systematic fashion.

2. If still in trouble, go to OPTIONS, and enter -1 for speed factor ('17') to try to get a clue while the computation progresses. Both the running concentration of each component (CC=) and its mass-balance residual (FF=) will be displayed.

3. Extremely bad points usually appear near inflections. There are no hard and fast rules, but one can try discarding them, adjusting millimoles of hydrogen present or redoing the experiment. When some refinement has been achieved (say SIGFIT ~.03 or less), check the pH difference column for egregious differences. These point to either bad entries or bad data. However larger differences are normal at steep inflections. If

the differences are systematic in a buffer region, model inadequacy is indicated in that a species may have been left out.

4. In handling data by computer, there is a tendency to refine all of the errors into the adjustable parameters, Beta's (or sometimes stoichiometric parameters) in this case. This is not advisable, but requires prudence in order to avoid. For these reasons millimole quantities, solution concentrations, volume, and other parameters should be determined prior to the computation of stability constants.

5. In order to refine 1:1 constants from 1:1 titration data and 2:1 constants from 2:1 data it is advantageous to use the computer operating system's resources on their lowest level: command(*.COM on the Vax) or batch (*.BAT on the PC) files. Thus in this example two files C1.COM and C2.COM would be set up as follows:

```
C1.COM (or C1.BAT)              C2.COM (or C2.BAT)
$Copy 1TO1.DAT FOR004.DAT       $Copy 2TO1.DAT FOR004.DAT
They would be used like this:
      @C1           prepare 1:1 data for input
      RUN BEST      start program
      .
      27            get current log betas
      .
      72            get out
      @C2           prepare 2:1 data for input
      RUN BEST      start program
      .
      27            get current log betas
      .
      72            get out
      @C1           start again
      .
      .            continue
```

This whole process could be automated using batch or command files so that intervention need not be required.

6. On the VAX one can redirect inputs from a batch file and run batch mode even on single calculations.

(Rev. 5

<u>DISKETTE</u>

Three files are provided for use with PROGRAM BEST:

BEST.FOR   Fortran source code

BEST.EXE   Compiled for running on a personal computer. If machine is incompatible, may have to recompile and relink the code.

IDAPB4.DAT   Sample input titration data to be copied into FOR004.DAT before running BEST.

```
C FIRST LINE OF BEST.FOR FILE
      PROGRAM BEST
C     VERSION (HP150 + VAX + PB)
      DOUBLE PRECISION NAME
      DIMENSION RCC(111,7),TITLE(20),VB(111),PH(111),NAME(7)
      DIMENSION AA(111,7),T(111,7)
      DIMENSION CC(7),CS(111),Y(7),PHCALC(111),A(7,7)
      DIMENSION D(7)
      REAL MMC(7),LBETA(111),MMACID,NB
      INTEGER*2 BELL
      INTEGER IS(111,7),IC(7)
      CHARACTER*2 NSTR
      COMMON /AINP/ SIGFIT,IS,IC,VB,PH,LBETA,MMC,NAME,TITLE
      COMMON /AAAA/ BELL,I1,I2,ISTEP,NDP,NS,NC,AA,RCC,T
      COMMON /BBBB/ PHCOR,V0,NB,MMACID
      COMMON /CCCC/ CC,CS,PHCALC
      COMMON /DDDD/ NOPT,FAST
      COMMON /EEEE/ A
      COMMON /GGGG/ NREF
      COMMON /LLLL/ LUNI,LUNO
      OPEN(1,FILE='FOR001.DAT',STATUS='UNKNOWN')
      OPEN(7,FILE='ERROR.DAT',STATUS='UNKNOWN')
      NREF=0
      BELL=7
C*****LUNI=0 FOR KEYBOARD ON HP150
C*****LUNO=0 FOR KEYBOARD ON HP150
C*****LUNI=5 FOR VAX TERMINALS
C*****LUNO=6 FOR VAX TERMINALS
      LUNI=0
      LUNO=0
      LUNI=5
      LUNO=6
      FAST=1.
      WRITE(LUNO,1) BELL
1     FORMAT(21X,'PROGRAM "BEST"',/
     X,21X,'RAMUNAS J. MOTEKAITIS'/
     1,21X,'VERSION 05/17/88',A4,/,
     2 21X,'[HP150] AND [VAX]',//)
      WRITE(LUNO,101)
101   FORMAT(//,' FILE USAGE:')
      WRITE(LUNO,102)
102   FORMAT(' FOR001.DAT - OUTPUT ',/,
     1' FOR002.DAT - SCRATCH (components) ',/,
     2' FOR003.DAT - SCRATCH (log beta values) ',/,
     3' FOR004.DAT - INPUT (original data)',/,
     4' ERROR .DAT - (logging OPTION 31)',/)
      NTH=0
9     FORMAT(I3)
      DO 88 K=1,7
      DO 88 I=1,111
88    RCC(I,K)=0.
      MM=0
27    CALL INPUT(MM,NTH)
      I1=1
      I2=NDP
      ISTEP=1
      CALL OPTION
      NDUM=1
      CALL BODY(NDUM)
      GO TO 98
45    CALL REFINE
      GO TO 98
36    CALL OUTPUT
98    WRITE(LUNO,77) BELL
77    FORMAT(/,1X,A4,'27=TOP   36=OUTPUT   45=REFINE   54=OPTIONS',
     1'  63=COMPUTE   72=STOP ? ',$)
      READ(LUNI,99) NSTR
99    FORMAT(A2)
      IF(NSTR.EQ.'27') GO TO 27
      IF(NSTR.EQ.'36') GO TO 36
      IF(NSTR.EQ.'45') GO TO 45
      IF(NSTR.EQ.'54') GO TO 54
      IF(NSTR.EQ.'63') GO TO 59
      IF(NSTR.EQ.'72') STOP
54    CALL OPTION
      GO TO 98
59    NDUM=-1
      CALL BODY(NDUM)
```

```
      GO TO 98
      END

      SUBROUTINE OPTION
      DOUBLE PRECISION NAME
      DIMENSION RCC(111,7),TITLE(20),VB(111),PH(111),NAME(7)
      DIMENSION AA(111,7),T(111,7)
      DIMENSION CC(7),CS(111),Y(7),PHCALC(111),A(7,7)
      DIMENSION D(7),BLER(111)
      REAL MMC(7),LBETA(111),MMACID,NB
      INTEGER IS(111,7),IC(7)
      INTEGER*2 BELL,SILENT
      CHARACTER*2 MCH
      COMMON /AINP/ SIGFIT,IS,IC,VB,PH,LBETA,MMC,NAME,TITLE
      COMMON /AAAA/ BELL,I1,I2,ISTEP,NDP,NS,NC,AA,RCC,T
      COMMON /BBBB/ PHCOR,V0,NB,MMACID
      COMMON /CCCC/ CC,CS,PHCALC
      COMMON /DDDD/ NOPT,FAST
      COMMON /EEEE/ A
      COMMON /LLLL/ LUNI,LUNO
      DATA SILENT/0/
      NOPT=1
      NTH=0
      GO TO 70
17    WRITE(LUNO,2) BELL
2     FORMAT(1X,A4,' ENTER SPEED FACTOR 1-5 or hit RETURN ',$)
      READ(LUNI,21) I
21    FORMAT(I2)
      IF(I.LT.0) NOPT=0
      IF(I.GT.6) NOPT=1
      IF(I.GT.0.AND.I.LT.6) FAST=10.0**I
      ISPEED=I
      GO TO 70
18    WRITE(LUNO,10) BELL
10    FORMAT(1X,A4,' INITIAL,FINAL(or -1),STEP (3I2)  or RETURN ',$)
      READ(LUNI,20) K1,K2,KSTEP
20    FORMAT(3I2)
      IF(K1.GT.0.AND.K1.LE.NDP) I1=K1
      IF(K2.GT.0.AND.K2.LE.NDP) I2=K2
      IF(KSTEP.GT.0) ISTEP=KSTEP
      IF(K2.LT.0) I2=NDP
      GO TO 70
59    DO 60 I=1,NS
60    WRITE(LUNO,47) I,LBETA(I),(IS(I,K),NAME(K),K=1,NC)
47    FORMAT(1X,I2,1X,F8.4,1X,8(I2,A9,1X))
48    WRITE(LUNO,49) BELL
49    FORMAT(1X,A4,'ENTER "I" OF LOG BETA TO BE CHANGED or RETURN ',$)
      READ(LUNI,51) I
51    FORMAT(I2)
      IF(I.LT.1) GO TO 70
      IF(I.GT.NS) GO TO 48
56    WRITE(LUNO,53) BELL,I
53    FORMAT(1X,A4,'ENTER NEW LOG BETA(',I2,')? ',$)
      READ(LUNI,57) LBETA(I)
57    FORMAT(F10.6)
      GO TO 59
70    WRITE(LUNO,68)
68    FORMAT('     OPTIONS:')
      WRITE(LUNO,69) ISPEED,I1,I2,ISTEP
69    FORMAT(5X,'14=REMOVE OR ADD DATA',/,5X,
     X'16=SILENT MODE(15=RESTORE)',/,5X,
     X'17=SPEED FACTOR(',I2,')',/,5X,
     X'18=POINTS(',3I3,')',/,5X,
     X'19=BETA VALUES')
      WRITE(LUNO,71) BELL,NB,MMACID,V0,PHCOR
71    FORMAT(1X,A4,1X,'20=MILLIMOLES',/,5X,
     X'21=NORMALITY BASE(',F9.5,')',/,5X,
     X'22=MILLIMOLES EXCESS ACID(',F9.5,')',/,5X,
     X'23=INITIAL VOLUME(',F9.5,')',/,5X,
     X'24 (RETURN)=END OF CHANGES ?',/,5X,
     X'25=PARABOLIC FUNCTION ?',/5X,
     X'26=APPORTION REMAINDER HYDROGEN TO EXCESS ACID ?',/,5X,
     X'27=READ CURRENT LOG BETA VALUES ?',/,5X,
     X'28=READ CURRENT MILLIMOLES ?',/,5X,
     X'29=CORRECTION ON PH CALCD(',F6.3,')? '/,5X
     X'30=AUTOREFINE MILLIMOLES ?'/,5X
     X'31=ERROR ANALYSIS ?',$)
      READ(LUNI,52) MCH
```

```
52    FORMAT(A2)
      IF(MCH.EQ.'  '.OR.MCH.EQ.'24'.OR.MCH.EQ.'63') GO TO 80
      IF(MCH.EQ.'14') GO TO 5300
      IF(MCH.EQ.'15') BELL=7
      IF(MCH.EQ.'16') BELL=SILENT
      IF(MCH.EQ.'17') GO TO 17
      IF(MCH.EQ.'18') GO TO 18
      IF(MCH.EQ.'19') GO TO 59
      IF(MCH.EQ.'20') GO TO 75
      IF(MCH.EQ.'21') GO TO 5225
      IF(MCH.EQ.'22') GO TO 5225
      IF(MCH.EQ.'23') GO TO 5225
      IF(MCH.EQ.'24') GO TO 80
      IF(MCH.EQ.'25') CALL PARB
      IF(MCH.EQ.'26') GO TO 5250
      IF(MCH.EQ.'27') GO TO 5215
      IF(MCH.EQ.'28') GO TO 5220
      IF(MCH.EQ.'29') GO TO 5225
      IF(MCH.EQ.'30') CALL STOICH(NC,NTH)
      IF(MCH.EQ.'31') CALL ERROR
      GO TO 70
5225  CONTINUE
5228  WRITE(LUNO,73) BELL
73    FORMAT(1X,A4,'NOW TYPE IN VALUE? ',$)
      IF(MCH.EQ.'21') READ(LUNI,57) NB
      IF(MCH.EQ.'22') READ(LUNI,57) MMACID
      IF(MCH.EQ.'23') READ(LUNI,57) V0
      IF(MCH.EQ.'29') READ(LUNI,57) PHCOR
      GO TO 70
75    WRITE(LUNO,76) (I,NAME(I),MMC(I),I=1,NC)
76    FORMAT(1X,I2,1X,A8,F10.5,3X,'MMOLES')
      WRITE(LUNO,77) BELL
77    FORMAT(1X,A4,' TYPE IN "I" ? or RETURN ',$)
58    READ(LUNI,581) SSI
581   FORMAT(F2.0)
      I=SSI
      IF(I.GT.NC) GO TO 75
      IF(I.LE.0) GO TO 70
      WRITE(LUNO,78) BELL,I
78    FORMAT(1X,A4,' TYPE IN MMC(',I2,')? ',$)
      READ(LUNI,57) MMC(I)
      GO TO 75
80    MM=1
      CALL INPUT(MM,NTH)
      MM=0
      RETURN
5215  OPEN(3,FILE='FOR003.DAT',STATUS='UNKNOWN')
      READ(3,5216) (LBETA(J),J=1,NS)
5216  FORMAT(F16.6)
      CLOSE(3)
      GO TO 5225
5220  OPEN(2,FILE='FOR002.DAT',STATUS='UNKNOWN')
      READ(2,5216) (MMC(J),J=1,NC),MMACID
      CLOSE(2)
      GO TO 5225
C APPORTION HYDROGEN TO MAKE SENSE OF A-VALUES
5250  WRITE(LUNO,5251) BELL
5251  FORMAT(1X,A4,'ENTER COMPONENT # AND MULTIPLIER (2I2) ?',$)
      READ(LUNI,20) NZC,NZM
      IF(NZC.LE.0.OR.NZC.GE.NC) GO TO 5225
      WRITE(LUNO,5252) BELL,NAME(NC),MMC(NC),MMACID
5252  FORMAT(1X,A4,A8,F10.5,'  AND EXCESS ACID ',F9.5,' CONVERTED TO')
      TOTMNC=MMC(NC)+MMACID
      MMC(NC)=NZM*MMC(NZC)
      IF(NZM.EQ.0) MMC(NC)=.00001
      MMACID=TOTMNC-MMC(NC)
      WRITE(LUNO,5253) NAME(NC),MMC(NC),MMACID
5253  FORMAT(1X,A10,F10.5,'  AND EXCESS ACID ',F9.5)
      GO TO 5225
C    ADD OR DROP A DATA POINT
5300  WRITE(LUNO,5301) BELL
5301  FORMAT(1X,A4,'"1"=ADD  "-1"=DROP A POINT',$)
      READ(LUNI,5306) NDA
      IF(NDA.EQ.-1) GO TO 53011
      IF(NDA.EQ.1) GO TO 53501
      GO TO 5300
53011 WRITE(LUNO,5302) BELL,(I,VB(I),PH(I),I=1,NDP)
5302  FORMAT(1X,A4,/,(5X,I10,2F8.3))
```

```
5303  WRITE(LUNO,5304) BELL
5304  FORMAT(1X,A4,
     x'REMOVE WHICH POINT("-1" RELIST  "0" RETURN)?',I2,$)
      READ(LUNI,5306) NREM
5306  FORMAT(I2)
      IF(NREM.EQ.0.OR.NREM.GT.NDP) GO TO 53528
      IF(NREM.LT.0) GO TO 53011
      NDP=NDP-1
      I2=I2-1
      DO 5308 I=NREM,NDP
      VB(I)=VB(I+1)
      PH(I)=PH(I+1)
5308  CONTINUE
      GO TO 53011
53501 WRITE(LUNO,5302) BELL,(I,VB(I),PH(I),I=1,NDP)
      WRITE(LUNO,5352) BELL
5352  FORMAT(1X,A4,
     x' ADD IN FRONT OF WHICH POINT("-1" RELIST  "0" RETURN) ?',$)
      READ(LUNI,5306) NADD
      IF(NADD.EQ.0.OR.NADD.GT.NDP) GO TO 53528
      IF(NADD.LT.0) GO TO 53501
53521 WRITE(LUNO,53522) BELL
53522 FORMAT(1X,A4,'VOLUME BASE ?',$)
      READ(LUNI,5216) VOLBAS
      NADDM=NADD-1
      IF(NADDM.EQ.0) NADDM=NADD
      IF(VOLBAS.EQ.VB(NADD).OR.VOLBAS.EQ.VB(NADDM)) GOTO 53521
53523 WRITE(LUNO,53524) BELL
53524 FORMAT(1X,A4,'-LOG (H+) ?',$)
      READ(LUNI,5216) PEEH
      IF(PEEH.EQ.PH(NADD).OR.PEEH.EQ.PH(NADDM)) GOTO 53523
      DO 53526 I=NADD,NDP
      II=NDP+NADD-I
      VB(II+1)=VB(II)
53526 PH(II+1)=PH(II)
      VB(NADD)=VOLBAS
      PH(NADD)=PEEH
      NDP=NDP+1
      I2=I2+1
      GO TO 53501
53528 MM=1
      CALL INPUT(MM,NTH)
      MM=0
      NDUM=-2
      CALL BODY(NDUM)
      GO TO 70
      END

      SUBROUTINE STOICH(NC,NTH)
      DOUBLE PRECISION X(3),Y(3),NAME
      DIMENSION VB(111),PH(111),NAME(7),TITLE(20),P(7)
      REAL MMC(7),LBETA(111)
      INTEGER*2 BELL
      CHARACTER*3 ADJ
      INTEGER IS(111,7),IC(7),MREF(7),FCT
      COMMON /AINP/ SIGFIT,IS,IC,VB,PH,LBETA,MMC,NAME,TITLE
      COMMON /FFFF/ X,Y,XX
      COMMON /GGGG/ NREF
      COMMON /LLLL/ LUNI,LUNO
      COMMON /MMMM/ ADJ
      DATA LIMCNT/5/,BELL/7/
      NCOUNT=0
      MM=1
      WRITE(LUNO,600) BELL
600   FORMAT(1X,A4,' REFINEMENT OF STOICHIOMETRY',/)
      WRITE(LUNO,6001)
6001  FORMAT(' CAUTION: Refinement of stoichiometric parameters'
     1' is usually',/,10X,'a highly questionable procedure and'
     2' should be avoided!',//)
      WRITE(LUNO,601)(I,NAME(I),MMC(I),I=1,NC)
601   FORMAT(1X,I2,1X,A8,F10.5,3X,' MMOLES')
      WRITE(LUNO,602)
602   FORMAT(1X,' COMPONENT NUMBERS(7I2) ?',$)
      READ(LUNI,603)(MREF(I),I=1,NC)
603   FORMAT(7I2)
      WRITE(LUNO,604) BELL
604   FORMAT(1X,A4,' ENTER SMALL INCREMENT',$)
      READ(LUNI,605) PP
```

```
605   FORMAT(F10.6)
      IF(PP.EQ.0.0) RETURN
      FCT=2
      WRITE(LUNO,610) BELL,LIMCNT
610   FORMAT(1X,A4,'ENTER MAX # ITERATIONS(default',1I3,')',$)
C*****NQUIT is a flag; 01 will FORTRAN STOP at end of this subroutine
      READ(LUNI,603) LIMCNT,NQUIT
      if(limcnt.eq.99)limcnt=999
      IF(LIMCNT.EQ.0) LIMCNT=5
      CALL INPUT(MM,NTH)
      NDUM=0
      CALL BODY(NDUM)
      CALL REFINE
      WRITE(LUNO,607)(MMC(I),I=1,NC),SIGFIT
      NREF=1
      DO 10 I=1,NC
10    P(I)=PP
100   S0=SIGFIT
105   J=0
110   S1=SIGFIT
      J=J+1
      M=MREF(J)
      IF(M.GT.0)GO TO 120
      GO TO 300
120   X(1)=MMC(M)
      Y(1)=S1
      MMC(M)=MMC(M)+P(J)
      CALL INPUT(MM,NTH)
      NDUM=0
      CALL BODY(NDUM)
      WRITE(LUNO,607)(MMC(I),I=1,NC),SIGFIT
      S2=SIGFIT
      X(2)=MMC(M)
      Y(2)=S2
      IF(S2.LE.S1) GO TO 130
      P(J)=-P(J)
      MMC(M)=MMC(M)+2*P(J)
      GO TO 140
130   MMC(M)=MMC(M)+P(J)
140   CALL INPUT(MM,NTH)
      NDUM=0
      CALL BODY(NDUM)
      WRITE(LUNO,609)(MMC(I),I=1,NC),SIGFIT
      S3=SIGFIT
      X(3)=MMC(M)
      Y(3)=S3
      CALL PARB
      Z1=ABS(MMC(M)-XX)
CCC   Factor FCT is variable now (is 2)
      Z2=ABS(P(J))*FCT
      IF(Z1.LE.Z2) GOTO150
      X(1)=X(2)
      Y(1)=Y(2)
      X(2)=X(3)
      Y(2)=Y(3)
      GO TO 130
150   MMC(M)=XX
      CALL INPUT(MM,NTH)
      NDUM=0
      CALL BODY(NDUM)
      WRITE(LUNO,609)(MMC(I),I=1,NC),SIGFIT
160   IF(J.LT.NC) GO TO 110
300   CONTINUE
      NCOUNT=NCOUNT+1
      WRITE(LUNO,606) NCOUNT
606   FORMAT(1X,'Refining Beta(s)****************',
     1'**********************************',I6)
C     WRITE(1,607)(MMC(I),I=1,NC),SIGFIT
      CALL REFINE
      S0MSF=S0-SIGFIT
      WRITE(LUNO,609)(MMC(I),I=1,NC),SIGFIT,S0MSF
609   FORMAT('+',8F10.6)
607   FORMAT(1X,8F10.6)
      IF(NCOUNT.GE.LIMCNT) GOTO 170
      IF((S0-SIGFIT).GE.0.0000005) GO TO 100
170   IF(NCOUNT.GE.LIMCNT) WRITE(LUNO,6071)
6071  FORMAT(' ITERATION LIMIT EXCEEDED')
      WRITE(LUNO,608) BELL
```

```
608   FORMAT(1X,A4,'*** END OF MMOLE REFINEMENT ***')
C ADJ="YES" IS A FLAG TO "OUTPUT" TO INDICATE "STOICH" WAS USED
      ADJ='YES'
      NREF=0
      IF(NQUIT.EQ.1) STOP
      RETURN
      END

      SUBROUTINE INPUT(MM,NTH)
      DOUBLE PRECISION NAME,RNAM
      DIMENSION RCC(111,7),TITLE(20),VB(111),PH(111),NAME(7)
      DIMENSION AA(111,7),T(111,7)
      DIMENSION CC(7),CS(111),Y(7),PHCALC(111),A(7,7)
      DIMENSION D(7),BLER(111)
      REAL MMC(7),LBETA(111),MMACID,NB
      INTEGER*2 BELL
      INTEGER IS(111,7),IC(7)
      CHARACTER*2 ECHO
      COMMON /AINP/ SIGFIT,IS,IC,VB,PH,LBETA,MMC,NAME,TITLE
      COMMON /AAAA/ BELL,I1,I2,ISTEP,NDP,NS,NC,AA,RCC,T
      COMMON /BBBB/ PHCOR,V0,NB,MMACID
      COMMON /CCCC/ CC,CS,PHCALC
      COMMON /DDDD/ NOPT,FAST
      COMMON /EEEE/ A
      COMMON /KKKK/ BLER
      COMMON /LLLL/ LUNI,LUNO
      DATA ECHO/'N'/
      IF(MM.EQ.1) GO TO 111
      OPEN(4,FILE='FOR004.DAT',STATUS='OLD')
      WRITE(LUNO,6610) BELL
6610  FORMAT(A4,' Do you want input file ECHO (Y/N) ?',$)
      READ(LUNI,6611) ECHO
6611  FORMAT(A1)
      IF(ECHO.EQ.'y') ECHO='Y'
      IF(ECHO.EQ.'Y') WRITE(LUNO,610)
610   FORMAT(' DISK INPUT FILE ECHO:')
      READ(4,601) (TITLE(I),I=1,20)
601   FORMAT(20A4)
      IF(ECHO.EQ.'Y') WRITE(LUNO,611) (TITLE(I),I=1,20)
611   FORMAT(1X,20A4)
      K=1
10    READ(4,602) RNAM,RMMC,RRCC
602   FORMAT(A8,3F8.5)
      IF(ECHO.EQ.'Y')WRITE(LUNO,612) RNAM,RMMC,RRCC
612   FORMAT(1X,A8,3F8.5)
      IF(RMMC.LE.0.) GO TO 12
      NAME(K)=RNAM
      MMC(K)=RMMC
      RCC(1,K)=RRCC
      IF(RCC(1,K).NE.0.) RCC(1,K)=10.0**RCC(1,K)
      K=K+1
      GO TO 10
12    NC=K-1
      NCM1=NC-1
      READ(4,502) V0,NB,MMACID,PHCOR
      IF(ECHO.EQ.'Y') WRITE(LUNO,503) V0,NB,MMACID,PHCOR
503   FORMAT(1X,9F8.4)
      I=1
100   READ(4,603) LBETA(I),(IC(K),IS(I,K),K=1,NC)
603   FORMAT(F8.5,14(2I2,1X))
      IF(ECHO.EQ.'Y')
     X WRITE(LUNO,613) LBETA(I),(IC(K),IS(I,K),K=1,NC)
613   FORMAT(1X,F9.5,14(2I2,1X))
C     INITIALIZING BLER ONLY WHEN FOR004.DAT IS READ OR REREAD
      BLER(I)=0
      IF(IC(1).EQ.0) GO TO 102
      I=I+1
      GO TO 100
102   NS=I-1
101   CONTINUE
C*****THEORETICAL DATA GENERATION   (not used)*****
      IF(NTH.LE.0) GO TO 105
      VB0=-.1
      PH0=1.9
      VB0=VB0+.1
      VB(I)=VB0
      PH0=PH0+.1
      PH(I)=PH0
```

```
112   CONTINUE
      NDP=NTH
      GO TO 111
105   I=1
104   READ(4,502,END=106) VB(I),PH(I),(RCC(I,K),K=1,NC)
      IF(ECHO.EQ.'Y')
     X WRITE(LUNO,502) VB(I),PH(I),(RCC(I,K),K=1,NC)
502   FORMAT(10F8.4)
      IF(RCC(1,NCM1).EQ.0.0) WRITE(LUNO,633) BELL
633   FORMAT(1X,A4,'FIRST DATA POINT NEEDS A GUESS VALUE')
      I=I+1
      GO TO 104
106   IF(ECHO.EQ.'Y') WRITE(LUNO,630)
630   FORMAT(' INPUT ECHO COMPLETE')
      NDP=I-1
      REWIND 4
      CLOSE (4)
111   DO 110 I=1,NDP
      V=V0+VB(I)
      CB=VB(I)*NB/V
      DO 109 K=1,NCM1
      IF(MM.EQ.0.AND.RCC(I,K).LT.0.) RCC(I,K)=10.0**RCC(I,K)
      T(I,K)=MMC(K)/V
109   AA(I,K)=CB/T(I,K)-MMACID/MMC(K)
      T(I,NC)=(MMC(NC)-NB*VB(I)+MMACID)/V
      AA(I,NC)=(CB*V-MMACID)/MMC(NC)
      IF(RCC(I,NC).LT.0.0) RCC(I,NC)=10.0**RCC(I,NC)
      IF(RCC(I,NC).EQ.0.0) RCC(I,NC)=10.0**(-PH(I))
110   CONTINUE
      RETURN
      END

      SUBROUTINE OUTPUT
      DOUBLE PRECISION NAME
      DIMENSION RCC(111,7),TITLE(20),VB(111),PH(111),NAME(7)
      DIMENSION AA(111,7),T(111,7)
      DIMENSION CC(7),CS(111),Y(7),PHCALC(111),A(7,7)
      DIMENSION CCL(7),D(7),BLER(111)
      DIMENSION NI(111),NEG(111)
      REAL MMC(7),LBETA(111),MMACID,NB
      INTEGER*2 BELL,FF
      INTEGER IS(111,7),IC(7)
      CHARACTER*1 DAT(9),TIM(8),AV
      CHARACTER*3 ADJ
      COMMON /AINP/ SIGFIT,IS,IC,VB,PH,LBETA,MMC,NAME,TITLE
      COMMON /AAAA/ BELL,I1,I2,ISTEP,NDP,NS,NC,AA,RCC,T
      COMMON /BBBB/ PHCOR,V0,NB,MMACID
      COMMON /CCCC/ CC,CS,PHCALC
      COMMON /DDDD/ NOPT,FAST
      COMMON /EEEE/ A
      COMMON /HHHH/ DLB,KCVERR,BETERR,SGF,BERBET
      COMMON /NEG/ NEGF,NEG
      COMMON /LLLL/ LUNI,LUNO
      COMMON /MMMM/ ADJ
      COMMON /KKKK/ BLER
      DATA FF/12/,AV/'A'/
      N1=I1
      N2=I2
      NSTEP=ISTEP
      DO 87 L=1,NS
87    NEG(L)=0
      IF(NEGF.EQ.1) GO TO 6181
      N=0
90    WRITE(LUNO,598) BELL
598   FORMAT(1X,A4,'5=TERMINAL   1=DISK   0=RETURN ? ',$)
      READ(LUNI,572) NPR
572   FORMAT(I3)
      IF(NPR.EQ.1.OR.NPR.EQ.5) GO TO 100
599   FORMAT(I2)
      IF(NPR.LE.0) RETURN
      GO TO 90
100   WRITE(LUNO,611) BELL
611   FORMAT(1X,A4,'1=DATA   2=A-VALUES   3=SPECIES   4=RESULTS ? ',$)
      READ(LUNI,599) NW
      IF(NPR.EQ.5) NPR=LUNO
      IF(NW.LT.1.OR.NW.GT.4) GO TO 90
      GO TO (11,22,33,44) NW
C*****INTRINSIC DATE AND TIME ROUTINES FOR VAX USED IN STATEMENT 11
```

```
CCCC11   CALL DATE(DAT)
CCCC     CALL TIME(TIM)
11    WRITE(NPR,600) FF, (TITLE(I),I=1,20)
600   FORMAT(A2,///,' Program BEST vers. 05/17/88',
     X/,10x,80A4/)
      WRITE(NPR,601) V0,NB,MMACID,NDP,PHCOR
601   FORMAT(10X,'INITIAL VOLUME',16X,F10.5,/,
     X  10X,'NORMALITY OF BASE',13X,F10.5,/,
     X  10X,'MILLIMOLES ACID',15X,F10.5,/,
     X  10X,'NUMBER DATA POINTS',12X,I10,/,
     X  10X,'PH CORR. INCLUDED',13X,F10.3,/)
      WRITE(NPR,602)
602   FORMAT(1X,' COMPONENTS:')
      WRITE(NPR,603) (K,NAME(K),MMC(K),K=1,NC)
603   FORMAT(1X,I9,A9,F11.5,'  MILLIMOLES')
      IF(ADJ.EQ.'YES'.AND.NPR.EQ.1) WRITE(NPR,6051)
6051  FORMAT('  NOTE: millimoles were adjusted')
      WRITE(NPR,604)
604   FORMAT(/,'  SPECIES:',/4X,'LOG BETA')
      DO 130 I=1,NS
130   WRITE(NPR,605)
     1 I,LBETA(I),(IS(I,K),NAME(K),K=1,NC)
605   FORMAT(1X,I2,1X,F8.4,1X,8(I2,A9,1X))
      WRITE(NPR,661) SIGFIT
C adjacent constant differences added to see if useful(5/09/86)
      WRITE(NPR,6052)
6052  FORMAT(/,' DIFFERENCE TABLE:("DIFF"=LOGBETA(I)-LOGBETA(I-1))',/
     X  1X,' I','      DIFF',5X,'ERROR')
      DLBETA=LBETA(1)
      DO 1301 I=1,NS
1302  IF(I.EQ.1) GO TO 1301
      DLBETA=LBETA(I)-LBETA(I-1)
1301  WRITE(NPR,6055) I,DLBETA,BLER(I)
6055  FORMAT(1X,I2,2F10.4)
      WRITE(NPR,608)
C  remove the above difference table lines if found unsatisfactory
      GO TO 100
22    WRITE(NPR,606) FF,(AV,K,K=1,NC)
606   FORMAT(A2,///,8X,'VB',7X,'PH',9(7X,A1,I1))
      DO 131 I=I1,I2,ISTEP
131   WRITE(NPR,607) VB(I),PH(I),(AA(I,K),K=1,NC)
607   FORMAT(1X,13F9.3)
      WRITE(NPR,608)
608   FORMAT(1X,/)
      GO TO 100
33    WRITE(LUNO,613) BELL
613   FORMAT(1X,A4,' 1=INDIVIDUAL SPECIES MODE',
     X' 2=PERCENTAGE SPECIES MODE ? ',$)
      DO 197 I=1,NS
197   NI(I)=0
200   READ(LUNI,599) NIP
      IF(NIP.LT.1.OR.NIP.GT.2) GO TO 33
      GO TO (210,220),NIP
210   WRITE(LUNO,609) BELL
609   FORMAT(1X,A4,' LIST INDIVIDUAL SPECIES')
      READ(LUNI,211) (NI(I),I=1,NS)
211   FORMAT(40I2)
220   WRITE(LUNO,614) BELL
614   FORMAT(1X,A4,' MINIMUM PER CENT ? ',$)
      READ(LUNI,615) WR
615   FORMAT(F8.3)
      WRP=WR/100.*T(1,1)
212   WRITE(LUNO,616) BELL
616   FORMAT(1X,A4,'N1,N2(-1),NSTEP? (3I2)? ',$)
      READ(LUNI,211)N1,N2,NSTEP
      IF(NSTEP.LE.0) NSTEP=1
      IF(N1.LT.I1.OR.N2.GT.I2) GO TO 212
      IF(N2.EQ.-1) N2=I2
      WRITE(NPR,618) FF
618   FORMAT(A2,///,3X,'#',1X,'LOG BETA',5X,'MOLARITY',6X,'%')
6181  DO 300 I=N1,N2,NSTEP
C  criterion for negligible species
      TEST=.000001*T(I,1)
      DO 240 K=1,NC
240   CCL(K)=ALOG10(RCC(I,K))
      DO 141 L=1,NS
      PL=0.
      DO 1403 K=1,NC
```

```
1403  PL=PL+CCL(K)*IS(L,K)
      CSL=PL+LBETA(L)
      CS(L)=10.0**CSL
C test for negligible species
      IF(NEG(L).EQ.1) GOTO 141
      IF(CS(L).GT.TEST) NEG(L)=1
141   CONTINUE
      IF(NEGF.EQ.1) GOTO 300
      WRITE(npr,6171) ph(i)
6171  FORMAT(' pH=',f7.3)
      DO 290 L=1,NS
      IF(NIP.EQ.2) GO TO 144
      DO 142 J=1,40
142   IF(L.EQ.NI(J)) GO TO 144
      GO TO 290
144   IF(CS(L).LT.WRP) GO TO 290
      PRO=CS(L)*100./T(I,1)
      WRITE(NPR,617) L,LBETA(L),CS(L),PRO,(IS(L,K),NAME(K),K=1,NC)
617   FORMAT(1X,I3,2X,F7.3,2X,1P,E11.3,0P,1X,F6.2,1X,7(I2,1X,A8,1X))
290   CONTINUE
300   CONTINUE
      IF(NEGF.EQ.1) RETURN
      GO TO 100
44    WRITE(NPR,630) FF
630   FORMAT(A2,///,15X,'VB',9X,'A',8X,'PH',4X,'PHCALC',6X,'DIFF')
      DO 288 I=I1,I2,ISTEP
      DD=0
      IF(PHCALC(I).NE.0) DD=PH(I)-PHCALC(I)
      WRITE(NPR,631) I,VB(I),AA(I,1),PH(I),PHCALC(I),DD
631   FORMAT(1X,I3,3X,5F10.3)
288   CONTINUE
      WRITE(NPR,661) SIGFIT
661   FORMAT(/,10X,'SIGMA PH FIT=',F10.6)
      GO TO 100
      END

      SUBROUTINE BODY(N)
      DOUBLE PRECISION NAME
      DIMENSION RCC(111,7),TITLE(20),VB(111),PH(111),NAME(7)
      DIMENSION AA(111,7),T(111,7)
      DIMENSION CC(7),CS(111),Y(7),PHCALC(111),A(7,7)
      DIMENSION FP(7,7),F(7),CCL(7),D(7)
      DIMENSION W(111)
      REAL MMC(7),LBETA(111),MMACID,NB
      INTEGER*2 BELL
      INTEGER IS(111,7),IC(7)
      COMMON /AINP/ SIGFIT,IS,IC,VB,PH,LBETA,MMC,NAME,TITLE
      COMMON /AAAA/ BELL,I1,I2,ISTEP,NDP,NS,NC,AA,RCC,T
      COMMON /BBBB/ PHCOR,V0,NB,MMACID
      COMMON /CCCC/ CC,CS,PHCALC
      COMMON /DDDD/ NOPT,FAST
      COMMON /EEEE/ A
      COMMON /LLLL/ LUNI,LUNO
      COMMON /HHHH/ DLB,KCVERR,BETERR,SGF,BERBET
      DATA NFIR/1/,SW/0.0/
C  -2 OR FORCES RECOMPUTATION OF WEIGHTING FACTORS FROM ADD DROP OPTION OR
'27'
      IF(N.EQ.-2.OR.N.EQ.1) NFIR=1
      IF(N.EQ.-2) N=-1
      IF(NFIR.EQ.0) GO TO 138
      NDPM1=NDP-1
C*******SQUARED WEIGHING FACTORS
      DO 1383 I=2,NDPM1
      W(I)=1/(PH(I+1)-PH(I-1))**2
1383  CONTINUE
C******.25 is 1/2**2
      W(1)=.25/(PH(2)-PH(1))**2
      W(NDP)=.25/(PH(NDP)-PH(NDPM1))**2
138   NFIR=0
      ITEND=30
      UNDER=10.**(-38.)
      NDPM1=-1
C****SET UP F(J)'S AND FP(J1,J2)'S****
      SIGFIT=0.0
      SW=0
      SNEG=T(NDP,1)*(1.0E-8)*FAST
      SLNEG=ALOG10(SNEG)
      I=I1-ISTEP
```

```
139   CONTINUE
      I=I+ISTEP
      SW=SW+W(I)
      IM1=I-ISTEP
      IM2=I-2*ISTEP
      IF(I.GT.I2) GO TO 200
1240  CONTINUE
1250  DO 1388 K=1,NC
      IF(RCC(I,K).LE.0.0) N=1
      IF(N) 1387,1387,1381
1381  IF(I.EQ.I1) CC(K)=RCC(I,K)
      IMSTEP=I-ISTEP
      IF(I.GT.I1) CC(K)=RCC(IMSTEP,K)
      IF(I.GT.(I1+ISTEP)) GO TO 1384
      GO TO 1388
1384  CALL NEXT(RCC(IM2,K),RCC(IM1,K),VB(IM2),VB(IM1),VB(I),CC(K))
      GO TO 1388
1387  CC(K)=RCC(I,K)
1388  CONTINUE
1260  DAMP=1.
1389  ITER=0
1390  ITER=ITER+1
      DO 1395 K=1,NC
1395  IF(CC(K).GT.0.0) CCL(K)=ALOG10(CC(K))
      DO 141 L=1,NS
      PL=0.0
      DO 1403 K=1,NC
1403  PL=PL+CCL(K)*IS(L,K)
      CSL=PL+LBETA(L)
      CS(L)=UNDER
      IF(CSL.GT.SLNEG) CS(L)=10.0**CSL
141   CONTINUE
      DO 150 J1=1,NC
      F(J1)=-T(I,J1)
      DO 150 J2=1,NC
150   FP(J1,J2)=0.
      DO 151 L=1,NS
      CT=CS(L)
      IF(CT.LT.SNEG) GO TO 151
      DO 1511 J1=1,NC
      IIS=IS(L,J1)
C     CHECK FOR ZERO COEFFICIENT
      IF(IIS.EQ.0) GO TO 1511
      F(J1)=F(J1)+IIS*CT
      DO 1512 J2=1,NC
      IISP=IIS*IS(L,J2)
      IF(IISP.EQ.0) GO TO 1512
      FP(J1,J2)=FP(J1,J2)+CT*IISP
1512  CONTINUE
1511  CONTINUE
151   CONTINUE
C***SIMULTANEOUS EQUATIONS SOLVED:***
      DO 160 J1=1,NC
      DO 160 J2=1,NC
CMULT:  CC(J2) ACTS AS COLUMN SCALING FACTOR
160   A(J1,J2)=FP(J1,J2)
      CALL GEL(DELTA,NC)
      IF(DELTA.EQ.0.0) DELTA=DELTAP
      IF(DELTA.NE.0.0) DELTAP=DELTA
      RDELTA=1./DELTA
      DO 164 K=1,NC
161   DO 162 J1=1,NC
      DO 162 J2=1,NC
CMULT:  CC(J2) ACTS AS A COLUMN SCALING FACTOR
162   A(J1,J2)=FP(J1,J2)
      DO 163 J1=1,NC
163   A(J1,K)=-F(J1)
      CALL GEL(D(K),NC)
164   CONTINUE
CMULT:  CC(K) ACTS AS A COLUMN SCALING FACTOR
      SUMQ=0.0
      SUMF=0.0
      DAMP=2.0
      DO 1650 K=1,NC
      IF(DAMP.LT.2.) DAMP=2.*DAMP
1642  DAMP=DAMP*.5
      R=D(K)/DELTA*DAMP*CC(K)
      Y(K)=CC(K)+R
```

```
      IF(Y(K).LE.0.0) GO TO 1642
      Q=ABS(R/Y(K))
      CC(K)=Y(K)
      SUMF=SUMF+ABS(F(K)/T(I,K))
1650  SUMQ=SUMQ+Q
      IF(NOPT.EQ.1) GO TO 1779
      ITEND=999
      WRITE(LUNO,1777) I,(CC(K),K=1,NC)
      WRITE(LUNO,1778) I,(F(K),K=1,NC)
1777  FORMAT(1X,' CC(',I2,')=',6E15.4)
1778  FORMAT(1X,'  F(',I2,')=',6E15.4/)
1779  IF(SUMQ.LE.1.0E-5.OR.SUMF.LT.1.E-4) GO TO 180
      IF(ITER.EQ.ITEND) GO TO 181
      GO TO 1390
180   IF(CC(NC).GT.0.) PHCALC(I)=-ALOG10(CC(NC))+PHCOR
      DO 179 K=1,NC
179   RCC(I,K)=CC(K)
      GO TO 183
181   PHCALC(I)=0
      IM1=I-ISTEP
      DO 182 K=1,NC
      IF(I.EQ.I1) GO TO 182
      RCC(I,K)=RCC(IM1,K)
182   CONTINUE
183   DIFF=0.
      IF(PHCALC(I).NE.0) DIFF=PH(I)-PHCALC(I)
      IF(N.EQ.-1.OR.N.EQ.1) WRITE(LUNO,620)
     X ITER,I,VB(I),PH(I),PHCALC(I),DIFF
620   FORMAT(' ',I3,3X,I3,3X,4F10.3)
      SIGFIT=SIGFIT+W(I)*DIFF*DIFF
      IF(PHCALC(I).NE.0.) NDPM1=NDPM1+1
      IF(I.GE.I2) GO TO 200
      GO TO 139
200   IF(NDPM1.LT.1) NDPM1=1
      IF(SW.EQ.0.)SW=1
      SIGFIT=SQRT(SIGFIT/SW)
      SGF=SIGFIT
      IF(N.EQ.-1.OR.N.EQ.1)  WRITE(LUNO,630) SIGFIT
630   FORMAT(1X,/,10X,'SIGMA PH FIT=',F10.6)
      RETURN
      END

      SUBROUTINE NEXT(Y1,Y2,X1,X2,X3,Y3)
      Y3=(Y2-Y1)*X3/(X2-X1)+(Y1*X2-Y2*X1)/(X2-X1)
      IF(Y3.LT.0) Y3=-Y3
      IF(Y3.EQ.0) Y3=Y2
      RETURN
      END

      SUBROUTINE GEL(DELTA,N)
      DIMENSION A(7,7)
      COMMON /EEEE/ A
      DELTA=1.
      DO 50 K=1,N
      AKK=A(K,K)
      IF(AKK.EQ.0.0) GO TO 21
      GO TO 41
21    DO 23 J=K,N
      IF(A(K,J).EQ.0.0) GO TO 23
      GO TO 31
23    CONTINUE
      DELTA=0.0
      GO TO 60
31    DO 34 I=K,N
      SAVE=A(I,J)
      A(I,J)=A(I,K)
34    A(I,K)=SAVE
C REDEFINE AKK
      AKK=A(K,K)
      DELTA=-DELTA
41    DELTA=DELTA*AKK
      IF((K-N).GE.0) GO TO 50
43    K1=K+1
      DO 46 I=K1,N
      C=A(I,K)/AKK
      DO 46 J=K1,N
46    A(I,J)=A(I,J)-C*A(K,J)
50    CONTINUE
```

```
60    CONTINUE
      RETURN
      END

      SUBROUTINE REFINE
      DOUBLE PRECISION NAME
      DIMENSION RCC(111,7),TITLE(20),VB(111),PH(111),NAME(7)
      DIMENSION AA(111,7),T(111,7)
      DIMENSION CC(7),CS(111),Y(7),PHCALC(111),A(7,7)
      DIMENSION D(7),NEG(111)
      DIMENSION KCV(20,111),U(20)
      REAL MMC(7),LBETA(111),MMACID,NB
      INTEGER*2 BELL
      INTEGER IS(111,7),IC(7)
      CHARACTER*2 STAR,BLANK,RF(111)
      COMMON /AINP/ SIGFIT,IS,IC,VB,PH,LBETA,MMC,NAME,TITLE
      COMMON /AAAA/ BELL,I1,I2,ISTEP,NDP,NS,NC,AA,RCC,T
      COMMON /BBBB/ PHCOR,V0,NB,MMACID
      COMMON /CCCC/ CC,CS,PHCALC
      COMMON /DDDD/ NOPT,FAST
      COMMON /EEEE/ A
      COMMON /GGGG/ NREF
      COMMON /HHHH/ DLB,KCVERR,BETERR,SGF,BERBET
      COMMON /NEG/ NEGF,NEG
      COMMON /LLLL/ LUNI,LUNO
      DATA IRLMT/5/
      DATA STAR/'* '/,BLANK/'  '/,RF/111*'  '/,KCV/2220*0/
      N=0
      NAUTO=0
      ITEMP=ITEMS
      IF(NREF.EQ.1) ITEM=99
      IF(NREF.EQ.1) GO TO 55
      WRITE(LUNO,51) BELL
51    FORMAT(1X,A4,
     X'CYCLES;ITEMS/CYCLE(99=repeat menu);[01]=ALL;[ITER LIMIT]'
     X,'(4I2)? ',$)
      READ(LUNI,605) NCY,ITEM,KSCR,IIRLMT
      IF(IIRLMT.GT.0.AND.IIRLMT.LT.19) IRLMT=IIRLMT
      IF(ITEM.LT.99) GO TO 54
      GO TO 55
54    ITEMS=ITEM
      DO 50 J=1,NC
      DO 50 K=1,NS
50    KCV(J,K)=0
55    IF(ITEM.EQ.99) ITEMS=ITEMP
      IF(NCY.EQ.0) NCY=1
      IF(ITEMS.EQ.0) ITEMS=1
52    CONTINUE
      N=N+1
      DO 505 J=1,ITEMS
      DF=1.0
      IR=1
      NSTOP=0
      U(IR)=SIGFIT
      SIGMIN=SIGFIT
      IF(NAUTO.EQ.1) GO TO 699
      IF(IIRLMT.EQ.-1) GO TO 6007
      IF(ITEM.EQ.99) GO TO 699
      DO 602 I=1,NS
602   WRITE(LUNO,603) I,LBETA(I),(IS(I,K),NAME(K),K=1,NC)
603   FORMAT(1X,I2,1X,F8.4,1X,8(I2,A9,1X))
      WRITE(LUNO,6033) SIGFIT
6033  FORMAT(10X,'SIGFIT=',F11.6)
      WRITE(LUNO,604) BELL
604   FORMAT(1X,A4,' ENTER "I" OF BETA AND COVARYING "I-S"(40I2)?')
      DO 680 JJ=1,ITEMS
6045  READ(LUNI,605) (KCV(JJ,I),I=1,NS)
605   FORMAT(40I2)
      IF(KCV(JJ,1).LE.0) RETURN
680   CONTINUE
C TWO STATS. ASSOCD. WITH SUBROUT. ERROR:
      KCVERR=KCV(1,1)
      BERBET=LBETA(KCVERR)
6007  WRITE(LUNO,606) BELL
606   FORMAT(1X,A4,' ENTER INCREMENT ? ',$)
      READ(LUNI,609) DLB
609   FORMAT(F10.4)
      IF(DLB.GT.2.0) DLB=2.0
```

```
      IF(DLB.LT.-2.0) DLB=-2.0
      IF(DLB.EQ.0) RETURN
699   IF(MSCR.EQ.-1) WRITE(LUNO,607)IR,SIGFIT,(LBETA(I),RF(I),I=1,NS)
608   FORMAT('0',1X,A4)
100   DO 700 I=1,NS
      II=KCV(J,I)
C     RF(II) is visual indicator for which constants are being refined
      IF(II.EQ.0) GO TO 700
      RF(II)=STAR
      LBETA(II)=LBETA(II)+DLB*DF
700   CONTINUE
      PSI=SIGFIT
      NDUM=0
      CALL BODY(NDUM)
      IF(SIGFIT.EQ.PSI) GO TO 500
      IR=IR+1
      IF(IR.GT.IRLMT) GO TO 500
      U(IR)=SIGFIT
      IF(KSCR.GT.0) WRITE(LUNO,607)
     X IR,SIGFIT,(LBETA(I),RF(I),I=1,NS)
607   FORMAT(1X,'SIG(',I2,')=',F8.6,2X,6(F8.4,A1),/,
     X (19X,F8.4,A1,F8.4,A1,
     1 F8.4,A1,F8.4,A1,F8.4,A1,F8.4,A1))
      DO 701 I=1,NS
701   RF(I)=BLANK
      IF(NSTOP.EQ.1) GO TO 400
190   CONTINUE
      IF(NSTOP.EQ.-3) GO TO 450
      IF(IR.EQ.2) GO TO 250
      IF(NSTOP.EQ.-2) GO TO 260
      IF(IR.LT.3) GO TO 100
195   CONTINUE
      IF(U(IR).LT.U(IR-1).AND.U(IR-1).LT.U(IR-2)) GO TO 100
      IF(U(IR).GT.U(IR-1).AND.U(IR-1).LT.U(IR-2)) GO TO 200
      IF(U(IR).GT.U(IR-1).AND.U(IR-1).GT.U(IR-2)) GO TO 300
      GO TO 100
200   X=-.5*(U(IR-2)-4.*U(IR-1)+3.*U(IR))/(U(IR-2)-2.*U(IR-1)+U(IR))
      DF=DF*X
      NSTOP=1
      IR=0
      GO TO 100
250   IF(U(1).GT.U(2)) GO TO 100
      HOLD=U(2)
      U(2)=U(1)
      U(1)=HOLD
      DF=-2.*DF
      NSTOP=-2
      GO TO 100
260   NSTOP =0
      DF=DF/2.
      GO TO 195
300   HOLD=U(IR)
      U(IR)=U(IR-2)
      U(IR-2)=HOLD
      DF=-3.*DF
      NSTOP=-3
      GO TO 100
400   DF=DF/X
      DF=.1*DF
      IF(ABS(DLB*DF).LT.0.101) GO TO 500
      NSTOP=0
      GO TO 190
450   NSTOP=0
      DF=DF/3.
      GO TO 195
500   CONTINUE
      NAUTO=1
505   CONTINUE
510   OPEN(3,FILE='FOR003.DAT',STATUS='UNKNOWN')
      WRITE(3,5216) (LBETA(I),I=1,NS)
5216  FORMAT(F16.6)
      CLOSE(3)
      OPEN(2,FILE='FOR002.DAT',STATUS='UNKNOWN')
      WRITE(2,5216) (MMC(J),J=1,NC),MMACID
      CLOSE(2)
      IF(NREF.EQ.1) GO TO 5249
      IF(MSCR.EQ.0)
     X WRITE(LUNO,607) IR,SIGFIT,(LBETA(I),RF(I),I=1,NS)
```

```
5249  IF(N.EQ.NCY) GOTO 5255
      GO TO 52
5255  IF(NREF.EQ.1) RETURN
C  OUTPUT routine determines neglible LBETAS
      NEGF=1
      CALL OUTPUT
      DO 5257 L=1,NS
      IF(NEG(L).EQ.0) WRITE(LUNO,5256)
     x L,LBETA(L),(IS(L,K),NAME(K),K=1,NC)
5256  FORMAT(' Less than 1 ppm:',1X,I2,1X,F8.4,1X,8(I2,A9,1X))
5257  CONTINUE
      NEGF=0
      RETURN
      END

      SUBROUTINE PARB
      DOUBLE PRECISION X(3),Y(3),TOP,BOT,A,B,C
      DIMENSION Z(7,7),G(3)
      REAL D,DELTA
      COMMON /EEEE/ Z
      COMMON /FFFF/ X,Y,XX
      COMMON /GGGG/ NREF
      COMMON /LLLL/ LUNI,LUNO
      IF(NREF.EQ.1) GO TO 210
      WRITE(LUNO,100)
100   FORMAT(' SUPPLY 3 PAIRS OF DATA, GET PARABOLIC MINIMUM')
      DO 200 I=1,3
      WRITE(LUNO,101) I
101   FORMAT(' ENTER PARAMETER',I2,' ? ',$)
      READ(LUNI,102) X(I)
102   FORMAT(D16.6)
      WRITE(LUNO,103) I
103   FORMAT(' ENTER SIGMA',I2,'     ? ',$)
      READ(LUNI,102) Y(I)
200   CONTINUE
210   A=Y(3)-Y(2)
      B=Y(3)-Y(1)
      C=Y(2)-Y(1)
      TOP=A*X(1)**2-B*X(2)**2+C*X(3)**2
      BOT=A*X(1)   -B*X(2)   +C*X(3)
      IF(BOT.EQ.0) GO TO 300
      XX=TOP/BOT/2.
C*****COMPUTE COEFFICIENTS ******
      K=-1
      N=3
209   K=K+1
      Z(1,1)=(X(1)-X(3))**2
      Z(2,1)=(X(2)-X(3))**2
      Z(3,1)=0
      Z(1,2)=X(1)-X(3)
      Z(2,2)=X(2)-X(3)
      Z(3,2)=0
      Z(1,3)=1.
      Z(2,3)=1.
      Z(3,3)=1.
      IF(K.EQ.0) CALL GEL(DELTA,N)
      IF(DELTA.EQ.0) GO TO 300
      IF(K.EQ.0) GO TO 209
      DO 214 I=1,3
214   Z(I,K)=Y(I)-Y(3)
      CALL GEL(D,N)
      G(K)=D/DELTA
      IF(K.LT.3) GO TO 209
      YY=G(1)*(XX-X(3))**2+G(2)*(XX-X(3))+G(3)+Y(3)
      IF(NREF.EQ.1) RETURN
      WRITE(LUNO,290) (X(I),Y(I),I=1,3),XX,YY
290   FORMAT(3(1X,2F16.6,/),/,' PARABOLIC MINIMUM=',F16.6,8X,
     X 'PREDICTED SIG=',F16.6,/)
      RETURN
300   WRITE(LUNO,305)
305   FORMAT(' CANNOT EVALUATE MINIMUM WITH PRESENT DATA')
      RETURN
      END

      SUBROUTINE ERROR
      DIMENSION BLER(111)
      INTEGER*2 BELL
      COMMON /GGGG/ NREF
```

```
      COMMON /HHHH/ DLB,KCVERR,BETERR,SGF,BERBET
      COMMON /KKKK/ BLER
      COMMON /LLLL/ LUNI,LUNO
      DATA NF/0/
C    DLB   :INCREMENT
C    BERBET:LOGBETA
C    BETERR:ERROR IN LOGBETA
C    KCVERR:WHICH SPECIES?
      IF(NF.EQ.1) GO TO 22
      BELL=7
      NF=1
      DO 11 I=1,111
11    BLER(I)=0.0
22    WRITE(LUNO,600) BELL
600   FORMAT(A4,' ERROR ANALYSIS DIRECTIONS:',/,
     1 '      -USE "01010101"',/,
     2 '      -USE BETA AND COVARYING BETAS',/,
     3 '      -USE INCREMENT IN PROPORTION TO SIGFIT SIZE',/,
     4 '      =RESULT WILL BE APPEAR IN FILE "ERROR.DAT"')
      SIG1=SGF
      NREF=0
      CALL REFINE
      SIG2=SGF
      NREF=1
      DLB=-DLB
      CALL REFINE
      NREF=0
      D=SIG2-SIG1
      IF(D.EQ.0) D=.0001
      BETERR=ABS(DLB*SIG1/D)
      WRITE(LUNO,605) KCVERR,BERBET,BETERR
605   FORMAT(' LOGBETA(',I2,')=',F10.4,'(',F10.4,')')
20    WRITE(7,605) KCVERR,BERBET,BETERR
      BLER(KCVERR)=BETERR
      RETURN
      END
C*******02/08/84-FP(J1,J2) Not calculated if IS(L,Ji)=0,i=1 or 2;
C*******03/21/84-SUBROUTINE STOICH ADDED
C*******-REFINE interval is a fixed quantity for DLB <= 1.00 now
C*******05/25/84-fixed T(1,1) used in WR
C*******05/26/84-SUBROUTINE INPUT: INPUT ECHO ADDED AT EACH READ
C*******06/05/84-OPTIONS REARRANGED
C*******10/09/84-FCT added to STOICH to allow change of extrapolation
C*******10/12/84-NEGligible species presence tests
C*******11/29/84-skip 5 lines on printing heading
C*******12/10/84-LIMCNT:a limit to number of iterations in STOICH
C*******04/30/85-APPORTION HYDROGEN BETWEEN MILLIMOLES AND EXCESS ACID
C*******07/12/85-WEIGHTING = 1/(DELTA pH)**2; negligible species output
C*******01/15/86-GEL IF(AKK.EQ.0) ==> IF(A(K,J).EQ.0); AKK=A(K,K) ADDED
C*******01/18/86-ITEND=999 WHEN DIAGNOSING
C*******01/25/86-#1779 IF(SUMQ.LT.1.E-5.)  from 1.E-4.
C*******03/14/86-DROP ADD POINTS ADDED IN OPTIONS
C*******03/24/86-FINISHED ADD POINTS IN OPTIONS; IRLMT=5 IN REFINE
C*******04/22/86-INPUT OF 7 COMPONENTS REPAIRED
C*******07/86-PDP11 CONVERTING TO HP150 MICROSOFT FORTRAN 77 CODE
C*******ERROR FUNCTION IS INTENDED TO BE USED WITHIN THE INTERVAL
C*******SENSITIVE TO THAT CONSTANT (pH RANGE = INTERVAL)
C*******05/17/88-COMPATIBLE FOR VAX AND MICROSOFT FORTRAN 77 CODE
C*******05/24/88-LUNI AND LUNO: LOGICAL UNIT #'S FOR INPUT AND OUTPUT; CALL (N)
C******* Last line (1153) of BEST.FOR
```

APPENDIX III

PROGRAM SPE

Program SPE is a FORTRAN computer program for the computation of species distributions from given stability constants (and, optionally, solubility products). The body of the program is extracted from the BEST program with the added subroutine PHASE which takes into account solubility products. The purpose of the program is two-fold:

1) To create an input file for a plotting program. In particular, the plotting program included is SPEPLOT.BAS which uses as input the output file FOR007.DAT from SPE and drives the Hewlett Packard LaserJet Series II or Plus$^+$ to make charts of % Species vs. p[H].

2) To create an optional FOR008.DAT output file containing a complete listing of each species, its stoichiometry, concentration, and relative percentage all as a function of p[H] spaced usually at 0.1 p[H] unit intervals in a span usually 10 p[H] units wide. The file is optional because it is usually very lengthy (and slows down program execution somewhat.) If FOR008.dat is selected it is usually not printed but conveniently perused within a word processor wherein searches and other commands may be given.

When solubility products are not specified, program execution is extremely rapid. The specification of solid phases slows down the execution because of the added computational overhead of checking, adding or subtracting matter. With extremely insoluble species, problems, if any, arise chiefly from fortran compiler imposed limitations on real number size.

The program was written and is intended to be a complementary utility to the visualization of solution composition as determined by known stability constants. SPE is particularly useful during work with BEST since their input files are virtually identical. Almost all of the illustrations of species distributions were generated with the help of SPE.FOR and SPEPLOT.BAS using either a Packard Bell 232 PC, an Hewlett Packard Model 150 II, a VAX, or a Micromation Z80 system drawn on the LaserJet$^+$ or Series II or on the Houston Instrument Hiplot

FILE USAGE

Program SPE requires one input file, produces one output file directed at plotting, and optionally writes a detailed species information file containing concentrations, and relative percentages as a function of p[H+]:

FOR001.DAT - input file (same as FOR004.DAT in BEST)
FOR007.DAT - output file (to be used by SPEPLOT)

FOR008.DAT - output file (for perusal or hard copy)

INPUT FILE

The name of the input file which SPE recognizes is FOR001.DAT (this name must be typed in on the supplied compiled version.) Its structure consists of four blocks which correspond to the first four blocks of the BEST input file.

BLOCK 1: one line of 80 characters of free format (80A1)

BLOCK 2: one line for each component name and quantity in millimoles - NAME(J) MMC(J) RCC(J) (8A1,2F8.5)

BLANK LINE: interpreted as end-of-component list

BLOCK 3: one blank line (i.e. V0 NB MMACID PHCOR (4F8.5) found in BEST is skipped over).

BLOCK 4: one line for each species log β, component number and stoichiometry - LBETA(K) IC(K) IS(I,K) (F8.5,14(2I2,1X))

BLANK LINE: interpreted as end-of-species list

(POSSIBLE SOLID)

The only subtle difference lay in BLOCK 2. Any iterative process requires an initial estimate of the answer. SPE uses a fixed fraction of the initial concentration of the components for its initial estimates. Sometimes these values are too far for successful convergence and they must be explicitly supplied. RCC(J) is the logarithm of the J'th component in its deprotonated form. Therefore, when SPE cannot generate good initial guesses internally, these should be supplied in the next 8 character field (1F8.3) adjacent to the components.

If the input file to BEST is used, SPE ignores any titration data present.

A solubility product is entered right after the blank line of BLOCK 4 (i.e. after the end-of-species list). It's format is the same as the soluble species format. When two or more solubility products are considered, they are entered on consecutive lines. The actual number of solubility products considered is given from the keyboard.

DIRECTIONS

1. Construct the input file FOR001.DAT according to above directions. One can use the BEST input file directly.
2. Run SPE and be prepared for an initial interactive session which sets up the desired running environment. The following questions will come up:

Query:	Response:
- a pause	RETURN
- INITIAL PH	RETURN or Specific p[H]
- FINAL PH	RETURN or Specific p[H]
- PH INCREMENT	RETURN or 0.1 or Interval for 100 points

- VOLUME	RETURN or Specific mL quantity
- OPTIONAl WRITE FILE	RETURN (NUL) or 8 for (FOR008.DAT)
- UNIT 1	(if asked, FOR001.DAT or filename)
- NO. SOLUBILITY PRODUCTS	RETURN (none) or enter number
- UNIT 8	(if asked, FOR008.DAT or filename)
- component & species table	
- SPECIES NOS. NOT TO PLOT	RETURN (best), eg. 1617 (16 and 17)
- ...DENOMINATOR PERCENTAGES	RETURN (1) or enter number
- MINIMUM %...WRITE FILE	RETURN (0) or enter small number
- UNIT 7	(if asked, FOR007.DAT)

Program now executes while writing to screen each pH value and rate of convergence factor SUMQ. SUMQ should normally be less than 1.E-4. When the ITERS is many the convergence factor may be larger.

Program terminates with the message "For printer plot RUN SPEPLOT.BAS"

3) Call up the BASIC Program SPEPLOT and follow its menus.

NOTES:

1) All Fortran real numbers require a decimal point while file numbers and species numbers require integers.
2) Use integral (decimal number but without fraction) values for pH.
3) Specify an PH INCREMENT consisting of 100 points per interval. For example pH 2 to 12 requires an increment of 0.1 unit; from 2-11 requires 0.09; etc.
4) The VOLUME is a convenient way of specifying the working concentration depending on the quantity of millimoles used in the input file.
5) Regarding NOS. NOT TO PLOT, the entries must be contiguous and include the last item. The usual usage is to easily eliminate the $[H^+]$ and $[OH^-]$ from the eventual plots. The example above assumes that species 16 and 17 are the last species in the list and correspond to H^+ and OH^- respectively. However, it is quite reasonable to hit RETURN and eliminate the unwanted species in the plotting program SPEPLOT.
6) Percentages computed for 1:1 systems can be calculated with respect to any component. For 2L:1M the percentages depend on which component is chosen. In SPE, the default

is Component #1, but provision is made to choose the denominator other than #1.

7) When solubility products are considered, certain defaults are assigned to the three convergence parameters CONV1 (0.0001), CONV2 (0.01) and CONV3 (0.0). CONV1 is the computation limit based on the rate of approach to self-consistency during removal or addition of matter. CONV2 is the tolerance within which the solubility products must be satisfied. CONV3 is a flag whereby any positive value will cause a voluminous diagnostic output to the screen indicating all the mass balance terms and actual solubility products during the computations. The idea is to use CONV3 to get a picture of the troublesome situation (if any) then to possibly relax the tolerance and convergence terms in order to get a speedier computation. The initial pause upon running SPE is actually the query for this triplet of convergence factors (3F10.6). Hitting RETURN here sets the default values.
8) While not mandatory, it is recommended to enter a ligand as the first component.

SPEPLOT.BAS

Although the program is written using Microsoft GWBASIC, no unusual machine specific extensions or functions (except TIME$) were employed making the program portable to systems using other BASIC's such as MBASIC. The program is menu-driven and self documenting.

SPEPLOT.BAS is complementary to SPE.FOR and uses the latter's output file FOR007.DAT as its input file. The input file contains a line of text as the header, followed respectively by initial pH, final pH, increment on pH, number of components, number of species, number of species to be plotted. The percentages are enumerated as a long list of repeating integers representing the percentages of species scaled by a factor of 10. The list is ended by a flag "-1", which is of course an impossible percentage. At the bottom of this file are tables of components and species. The top and bottom portions of FOR007.DAT are illustrated with data taken from an Oxalic acid calculation in the absence of metal ions.:

```
oxalic acid page 95 5/24/88
   2.0000   12.1000     .1000
  3   5   2
911
 74
 14
922
 59
 17
929
 47
 22
```

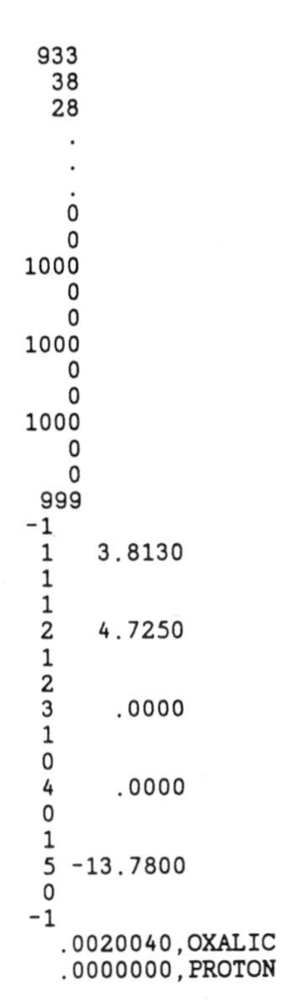

```
933
 38
 28
  .
  .
  .
  0
  0
1000
  0
  0
1000
  0
  0
1000
  0
  0
999
-1
1   3.8130
1
1
2   4.7250
1
2
3    .0000
1
0
4    .0000
0
1
5 -13.7800
0
-1
 .0020040,OXALIC
 .0000000,PROTON
```

The numbers indicate pH from 2.0 to 12.1 in steps of 0.1, 5 species present of which 3 are plotted and the number of components present is 2. The triads of similar numbers represent percents multiplied by ten ending in a first negative number -1. Next come species number, log B, stoichiometric coefficients for all 5 species present ending in a flag (-1, impossible species number). The final two entries are the component concentrations with the names.

<u>DISKETTE</u>

Three files are provided for use with PROGRAM SPE:

- SPE.FOR — Fortran source code
- SPE.EXE — Compiled for running on a personal computer. If machine is incompatible, may have to recompile SPE.FOR and relink.
- SPEPLOT.BAS — Plotting Program written using GWBASIC for use with LASERJET$^+$ or Series II. Generate FOR007.DAT file before running SPEPLOT.

In addition to program specific files, information how to get technical support may be found in the file HELPME.DOC which could be simply typed on the terminal.

```
         PROGRAM SPE
         DIMENSION TMEM(9)
         CHARACTER*80 TITLE
         COMMON /AAAA/ PH,PHI,PHF,PHINC
         COMMON /EEEE/ TITLE
         COMMON /GGGG/ NPRINT,NSP,N1,N2
         COMMON /IIII/ TMEM,ADJ
         COMMON /TEST/ CONV1,CONV2,CONV3
         READ(*,303) CONV1,CONV2,CONV3
303      FORMAT(3F10.6)
         WRITE(*,10)
10       FORMAT(/////,' SPECIES DISTRIBUTION PROGRAM',//)
         WRITE(*,20)
20       FORMAT(' SPE.FOR vs. 08/18-24/87 PC, hp 150 and vax',//)
         WRITE(*,30)
30       FORMAT(' Ramunas J. Motekaitis',//)
         WRITE(*,40)
40       FORMAT(
     1   ' PROGRAM FILES -',/,
     2   '      INPUT:          UNIT 1',/,
     3   '      PLOTFILE:       UNIT 7',/,
     2   '      WRITE:          UNIT 8 (OPTIONAL)'/)
         CALL INPUT
700      ADJ=0
         CALL SOLVE
         IF(NSP.EQ.0) GO TO 800
         CALL PHASE
800      PH=PH+PHINC
C*******REMOVE C'S FROM COLUMN 1 TO ENABLE MANUAL INPUT:
C         WRITE(*,2)
C2        FORMAT(' ENTER PH')
C         READ(*,1) PH
C1        FORMAT(F10.3)

         IF(PH.LE.PHF) GO TO 700
100      CONTINUE
         WRITE(*,200)
200      FORMAT(1X,' For printer plot run SPEPLOT.BAS')
         END
C********
         SUBROUTINE INPUT
         DIMENSION CC(9),CCL(9),CS(81),F(9),FP(9,9),A(9,9)
         DOUBLE PRECISION NAME(9)
         DIMENSION NOT(81)
         DIMENSION BIG(81),PHBIG(81)
         REAL MMC(9),LBETA(81)
         DIMENSION T(9),D(9),RCC(9),IC(81,9),IS(81,9)
         DIMENSION TMEM(9)
         CHARACTER*80 TITLE
         COMMON /AAAA/ PH,PHI,PHF,PHINC
         COMMON /BBBB/ LBETA,NAME,NP,WR
         COMMON /CCCC/ CC,CCL,CS,F,FP,T,D,NC,NS,NCM1,IC,IS,NOT
         COMMON /EEEE/ TITLE
         COMMON /FFFF/ A
         COMMON /GGGG/ NPRINT,NSP,N1,N2
         COMMON /HHHH/ BIG,PHBIG
         COMMON /IIII/ TMEM,ADJ
         DO 400 I=1,80
         BIG(I)=0
400      PHBIG(I)=0
         WRITE(*,6012)
6012      FORMAT(
     1   ' PROGRAM DEFAULTS:',/,
     2   '      INITIAL PH:      2.00',/,
     3   '      FINAL   PH:     10.00 + INITIAL VALUE',/,
     4   '      PH INCREMENT:    0.10',/,
     5   '      VOLUME (ML):    50.00',/,
     6   '      DENOM. COMPONENT:    1')

         WRITE(*,600)
600      FORMAT(1X,' INITIAL PH ?',\)
         READ(*,700) PHI
700      FORMAT(F10.3)
         WRITE(*,601)
601      FORMAT(1X,' FINAL PH ?',\)
         READ(*,700) PHF
         WRITE(*,602)
602      FORMAT(1X,' PH INCREMENT ?',\)
```

```
      READ(*,700) PHINC
      WRITE(*,6020)
6020  FORMAT(' VOLUME ?',\)
      READ(*,700) VOL
6029  WRITE(*,603)
603   FORMAT(1X,' OPTIONAL WRITE FILE NUMBER ?',\)
      READ(*,501) NPRINT
501   FORMAT(2I2)
      NOPT=1
      IF(NPRINT.EQ.1.OR.NPRINT.EQ.7) GO TO 6029
      READ(NOPT,499) TITLE
499   FORMAT(A)
      IF(PHI.EQ.0) PHI=2.
      IF(PHF.EQ.0) PHF=PHI+10.1
      IF(PHINC.EQ.0) PHINC=.1
      IF(VOL.EQ.0) VOL=50.
      PH=PHI
      K=1
10    READ(NOPT,502) NAME(K),MMC(K), RCC(K)
      IF(MMC(K).EQ.0) GO TO 12
      T(K)=MMC(K)/VOL
      TMEM(K)=T(K)
      IF(RCC(K).EQ.0) RCC(K)=ALOG10(T(K))-6.0
      CC(K)=10.0**RCC(K)
      K=K+1
      GO TO 10
12    NC=K-1
      T(NC)=0
      TMEM(NC)=0
      NCM1=NC-1
502   FORMAT(A8,2F8.5)
      READ(NOPT,506)DUMMY
506   FORMAT(F8.4)
      I=1
100   READ(NOPT,503) LBETA(I),(IC(I,K),IS(I,K),K=1,NC)
503   FORMAT(F8.5,14(2I2,1X))
C  ALUMINUM POLYHYDROXIDE SPECIAL:
      IF(IS(I,NC).EQ.32) IS(I,NC)=-IS(I,NC)
C  REMOVE THE ABOVE STATEMENT IF IT OFFENDS YOU
      IF(IC(I,1).EQ.0) GO TO 102
      I=I+1
      GO TO 100
102   NS=I-1
      NSPRNT=NS
C******** SOLUBILITY CONSIDERATION
      WRITE(*,5030)
5030  FORMAT(' NUMBER OF SOLUBILITY PRODUCTS?',\)
      READ(*,501) NSP
      IF(NSP.LE.0) GO TO 1021
      N1=NS+1
      N2=NS+NSP
      DO 5031 I=N1,N2
5031  READ(NOPT,503) LBETA(I),(IC(I,K),IS(I,K),K=1,NC)
C********INPUT ECHO
      IF(NPRINT.GT.0) WRITE(NPRINT,4991) TITLE
1021  WRITE(*,4991) TITLE
4991  FORMAT(1X,A)
      WRITE(*,3002)
      IF(NPRINT.GT.0) WRITE(NPRINT,3002)
3002  FORMAT(1X, ' COMPONENTS:')
      WRITE(*,613) (K,NAME(K),MMC(K),T(K),K=1,NC)
      IF(NPRINT.GT.0) WRITE(NPRINT,613) (K,NAME(K),MMC(K),T(K),K=1,NC)
613   FORMAT(0P,I9,A9,F11.5,' MILLIMOLES',8X,1P,E11.2,' MOL/LITER')
      WRITE(*,604)
604   FORMAT(1H0,' SPECIES:',/,4X,'LOG BETA',3X,'MAX %     @ pH')
      IF(NPRINT.GT.0) WRITE(NPRINT,604)
      DO 130 I=1,NS
      IF(NPRINT.GT.0)
     1 WRITE(NPRINT,593) I,LBETA(I),(IS(I,K),NAME(K), K=1,NC)
130   WRITE(*,593) I,LBETA(I),(IS(I,K),NAME(K),K=1,NC)
593   FORMAT(1X,0P,I2,1X,F8.4,1X,8(I3,A8,1X))
      IF(NSP.EQ.0) GO TO 6045
      WRITE(*,5931)
      IF(NPRINT.GT.0) WRITE(NPRINT,5931)
5931  FORMAT(' SOLUBILITY PRODUCTS:')
      DO 1301 I=N1,N2
      IF(NPRINT.GT.0)
     1 WRITE(NPRINT,593) I,LBETA(I),(IS(I,K),NAME(K), K=1,NC)
```

```
1301    WRITE(*,593) I,LBETA(I),(IS(I,K),NAME(K),K=1,NC)
6045    I1=1
        I2=40
        IF(NS.LE.40) I2=NS
6052    WRITE(*,6050) I1,I2
6050    FORMAT(' ENTER SPECIES #S NOT TO PLOT (',I2,'-',I2,')',\)
        READ(*,6051) (NOT(I),I=I1,I2)
6051    FORMAT(40I2)
        I1=I1+40
        I2=I2+40
        IF(I1.GT.NS) GO TO 6153
        IF(I2.GT.NS) I2=NS
        IF(I2.LE.NS) GO TO 6052
6153    IF(NOT(1).GE.0) GO TO 6154
        NOT(1)=NS
        NOT(2)=NS-1
6154    IF(NPRINT.GT.0) WRITE(NPRINT,6155) (NOT(I),I=1,NS)
6155    FORMAT(' SPECIES DELETED FROM GRAPH:',/,(5X,10I4))
        WRITE(*,6053)
6053    FORMAT(' ENTER COMPONENT NUMBER AS DENOMINATOR FOR',
     1 ' PERCENTAGES ?',\)
        READ(*,504) NP
504     FORMAT(I1)
        IF(NP.EQ.0.OR.NP.GE.NC) NP=1
        WRITE(*,6054)
6054    FORMAT(' ENTER MINIMUM % FOR OPTIONAL WRITE FILE ?',\)
        READ(*,700) WR
C****** FOR PLOTTER INFO
        WRITE(7,499) TITLE
        WRITE(7,6055) PHI,PHF,PHINC
6055    FORMAT(3F10.4)
        DO 1001 ISP=1,NS
1001    IF(NOT(ISP).GT.0) NSPRNT=NSPRNT-1
        NSPRNT=NSPRNT+NSP
        NST=NS+NSP
        WRITE(7,1002) NSPRNT,NST,NC
1002    FORMAT(3I4)
        RETURN
        END
C********
        SUBROUTINE SOLVE
        DIMENSION CC(9),CCL(9),CS(81),F(9),FP(9,9),A(9,9)
        DOUBLE PRECISION NAME(9)
        DIMENSION NOT(81)
        REAL MMC(9),LBETA(81)
        DIMENSION T(9),D(9),RCC(9),IC(81,9),IS(81,9)
        DIMENSION TMEM(9)
        DIMENSION BIG(81),PHBIG(81)
        CHARACTER*80 TITLE
        COMMON /AAAA/ PH,PHI,PHF,PHINC
        COMMON /BBBB/ LBETA,NAME,NP,WR
        COMMON /CCCC/ CC,CCL,CS,F,FP,T,D,NC,NS,NCM1,IC,IS,NOT
        COMMON /EEEE/ TITLE
        COMMON /FFFF/ A
        COMMON /GGGG/ NPRINT,NSP,N1,N2
        COMMON /HHHH/ BIG,PHBIG
        COMMON /IIII/ TMEM,ADJ
        DATA ITEND/50/,DAMP/1.0/,II/0/
        DATA NFIRST/1/,UNDER/1.E-38/
        IF(NFIRST.EQ.0) GO TO 1389
        SNEG=T(1)*(1.E-8)
        SLNEG=ALOG10(SNEG)
        TNPR=100./TMEM(NP)
        NFIRST=0
1389    ITER=0
        CC(NC)=10.**(-PH)
1390    ITER=ITER+1
        DO 1395 K=1,NC
1395    CCL(K)=ALOG10(CC(K))
        DO 141 L=1,NS
        PL=0
        DO 1403 K=1,NC
1403    PL=PL+CCL(K)*IS(L,K)
        CSL=PL+LBETA(L)
        CS(L)=UNDER
        IF(CSL.GT.(-38.)) CS(L)=10.**CSL
141     CONTINUE
        DO 150 J1=1,NCM1
```

```
        F(J1)=-T(J1)
        DO 150 J2=1,NCM1
150     FP(J1,J2)=0
        DO 151 L=1,NS
        CT=CS(L)
        IF(CT.LT.SNEG)  GO TO 151
        DO 1511 J1=1,NCM1
        IIS=IS(L,J1)
        IF(IIS.EQ.0) GO TO 1511
        F(J1)=F(J1)+IIS*CT
        DO 1512 J2=1,NCM1
        IISP=IIS*IS(L,J2)
        IF(IISP.EQ.0) GO TO 1512
        FP(J1,J2)=FP(J1,J2)+CT*IISP
1512    CONTINUE
1511    CONTINUE
151     CONTINUE
C       SOLUTION OF SIMULTANEOUS EQUATIONS
        DO 160 J1=1,NCM1
        DO 160 J2=1,NCM1
160     A(J1,J2)=FP(J1,J2)
      IF(DELTA.NE.0.0) DELTAP=DELTA
        CALL GEL(DELTA,NCM1)
      IF(DELTA.EQ.0.0) DELTA=DELTAP
        DO 164 K=1,NCM1
        DO 162 J1=1, NCM1
        DO 162 J2=1,NCM1
162     A(J1,J2)=FP(J1,J2)
        DO 163 J1=1,NCM1
163     A(J1,K)=-F(J1)
        CALL GEL(D(K),NCM1)
164     CONTINUE
        SUMQ=0
        SUMF=0
        DAMP=2.
        DO 1650 K=1,NCM1
        IF(DAMP.LT.2.) DAMP=DAMP*2
1642    DAMP=DAMP*.5
        R=D(K)/DELTA*CC(K)*DAMP
        YK=CC(K)+R
        IF(YK.LE.0.) GO TO 1642
        Q=ABS(R/YK)
        CC(K)=YK
        SUMF=SUMF+ABS(F(K)/T(K))
1650    SUMQ=SUMQ+Q
        IF(ITER.EQ.1) GO TO 1390
        IF(ITER.GT.ITEND) GO TO 180
        IF(SUMQ.LT.1.E-4.OR.SUMF.LT.1.E-5) GO TO 180
        GO TO 1390
180     CONTINUE
        IF(NSP.EQ.0) GO TO 181
        IF(ADJ.NE.-1) GO TO 615
181     IF(NPRINT.GT.0) WRITE(NPRINT,605) PH,SUMQ,ITER
        WRITE(*,605) PH,SUMQ,ITER
605     FORMAT(' PH= ',F6.3,10X,'SUMQ= ',1P,E11.1,I3,' ITERS')
        NSNSP=NS+NSP
        DO 190 L=1,NSNSP
        DO 6050 J=1,NSNSP
6050    IF(NOT(J).EQ.L) GO TO 190
        PRO=ABS(CS(L)*TNPR)
        IF(PRO.LT.WR) GO TO 189
        IF(NPRINT.LT.0)WRITE(*,610)
     X      LBETA(L),CS(L),PRO,(IS(L,K),NAME(K),K=1,NC)
        IF(NPRINT.GT.0) WRITE(NPRINT,610)
     X      LBETA(L),CS(L),PRO,(IS(L,K),NAME(K),K=1,NC)
610     FORMAT(1X,F7.3,2X,1P,E11.3,0P,1X,F6.2,1X,7(I3,A8,1X))
189     NPRO=PRO*10
4       WRITE(7,6064) NPRO
6064    FORMAT(I5)
190     CONTINUE
        IF(PH+PHINC.LE.PHF) GO TO 615
        WRITE(7,619)
619     FORMAT(' -1')
        DO 614 L=1,NSNSP
        WRITE(7,611) L,LBETA(L),(IS(L,K),K=1,NC)
611     FORMAT(I3,F9.4,/,(I3))
614     CONTINUE
        T(NC)=0
```

```
        TMEM(NC)=0
        WRITE(7,6149) (TMEM(K),NAME(K),K=1,NC)
6149    FORMAT(1X,F10.7,',',A8)
        WRITE(7,611) NSP
615     RETURN
        END
C********
        SUBROUTINE GEL(DELTA,N)
        DIMENSION A(9,9)
        COMMON /FFFF/ A
        DELTA=1
        DO 50 K=1,N
        AKK=A(K,K)
        IF(AKK.EQ.0) GO TO 21
        GO TO 41
21      DO 23 J=K,N
        IF(A(K,J).EQ.0) GO TO 23
        GO TO 31
23      CONTINUE
        DELTA=0
        GO TO 60
31      DO 34 I=K,N
        SAVE=A(I,J)
        A(I,J)=A(I,K)
34      A(I,K)=SAVE
        AKK=A(K,K)
        DELTA=-DELTA
41      DELTA=DELTA*AKK
        IF((K-N).GE.0) GO TO 50
43      K1=K+1
        DO 46 I=K1,N
        C=A(I,K)/AKK
        DO 46 J=K1,N
46      A(I,J)=A(I,J)-C*A(K,J)
50      CONTINUE
60      CONTINUE
        RETURN
        END
C
C       PHASE.FOR subroutine to be used with SPE.for
C       Fortran 77 compatible  runs on hp150 and vax (06/02/88)
C       08/26/87  WORKING COPY
        SUBROUTINE PHASE
        DIMENSION CC(9),CCL(9),CS(81),F(9),FP(9,9),A(9,9)
        DOUBLE PRECISION NAME(9)
        DIMENSION NOT(81)
        DIMENSION BIG(81),PHBIG(81)
        REAL MMC(9),LBETA(81)
        INTEGER PPT,PASS
        DIMENSION BIN(81,9)
        DIMENSION T(9),D(9),RCC(9),IC(81,9),IS(81,9)
        DIMENSION TMEM(9)
        CHARACTER*80 TITLE
        COMMON /AAAA/ PH,PHI,PHF,PHINC
        COMMON /BBBB/ LBETA,NAME,NP,WR
        COMMON /CCCC/ CC,CCL,CS,F,FP,T,D,NC,NS,NCM1,IC,IS,NOT
        COMMON /EEEE/ TITLE
        COMMON /FFFF/ A
        COMMON /GGGG/ NPRINT,NSP,N1,N2
        COMMON /HHHH/ BIG,PHBIG
        COMMON /IIII/ TMEM,ADJ
        COMMON /TEST/ CONV1,CONV2,CONV3
        DATA BIN/729*0./
        IF(CONV1.EQ.0) CONV1=.0001
        IF(CONV2.EQ.0) CONV2=.01
        IF(CONV3.EQ.0) CONV3=0.0
c debug         WRITE(*,33)
33      FORMAT(' SUBROUTINE PHASE')
        PASS=0
C ADJ=1 WHEN ANY T IS ADJUSTED
1803    ADJ=0
        PASS=PASS+1
        DO 1800 L=N1,N2
C SOLUBILITY PRODUCT
1805    PL=0
        DO 1810 K=1,NC
1810    PL=PL+CCL(K)*IS(L,K)
      PLDIF=PL-LBETA(L)
```

```
        IF(PL.GT.LBETA(L)) GO TO 1811
C CHECK IF ANYTHING CAN BE DISSOLVED
      PPT=0
      DO 18103 K=1,NCM1
      IF(IS(L,K).EQ.0) GO TO 18103
      IF(T(K).LT.TMEM(K)) PPT=1
18103      CONTINUE
      IF(PPT.EQ.0) GO TO 1800
C PROCEED WITH DISSOLVING SOLID PHASE(S)
      RAT=.1
      RAT=-PLDIF*.1
18104      DO 18105 K=1,NCM1
      IF(IS(L,K).EQ.0.0) GO TO 18105
      TR=T(K)*IS(L,K)*RAT
      IF((BIN(L,K)-TR).LT.0.0) TR=BIN(L,K)
18105      CONTINUE
      DO 18107 K=1,NCM1
      IF(IS(L,K).EQ.0.0) GO TO 18107
      T(K)=T(K)+TR
        BIN(L,K)=BIN(L,K)-TR
        IF(CONV3.GT.0.0) WRITE(*,55) L,K,IS(L,K),PL,LBETA(L),RAT,
     1    (T(J),J=1,NCM1),(BIN(L,J),J=1,NCM1)
18107      CONTINUE
      CALL SOLVE
      PL=0
      DO 18108 K=1,NC
18108      PL=PL+CCL(K)*IS(L,K)
      IF(TR.EQ.0.0) GO TO 1800
      IF(PL.LT.LBETA(L).AND.RAT.GT.0.0) GO TO 18104
      IF(PL.GT.LBETA(L).AND.RAT.LT.0.0) GO TO 18104
      IF(ABS(RAT).LT.CONV1) GO TO 1800
      RAT=-RAT*.1
      GO TO 18104

C REMOVE MATTER FROM SOLUTION
1811      RAT=0.1
        RAT=PLDIF*.1
1813    DO 1815 K=1,NCM1
      TR=T(K)*IS(L,K)*RAT
      T(K)=T(K)-TR
        BIN(L,K)=BIN(L,K)+TR
        IF(CONV3.GT.0.0) WRITE(*,55) L,K,IS(L,K),PL,LBETA(L),RAT,
     1    (T(J),J=1,NCM1),(BIN(L,J),J=1,NCM1)
55      FORMAT(1X,3I3,1P,(6E11.4))
1815    CONTINUE
1816    CALL SOLVE
1817    PL=0
        DO 1818 K=1,NC
1818    PL=PL+CCL(K)*IS(L,K)
        IF(PL.GT.LBETA(L).AND.RAT.GT.0.0) GO TO 1813
      IF(PL.LT.LBETA(L).AND.RAT.LT.0.0) GO TO 1813
      IF(ABS(RAT).LT.CONV1) GO TO 1800
      RAT=-RAT*0.1
      GO TO 1813
1800    CONTINUE
        CALL SOLVE

C  CHECK IF ALL Ksp ARE SATISFIED WITHIN "CONV2" TOLERANCE
      PPT=0
      DO 1830 L=N1,N2
      PL=0
        DO 1831 K=1,NC
1831    PL=PL+CCL(K)*IS(L,K)
      IF((PL-LBETA(L)).LE.CONV2) GO TO 1830
      PPT=1
1830      CONTINUE
      WRITE(*,1832) PASS
1832      FORMAT(' PASS=',I5)
      IF(PASS.GT.10) WRITE(*,1833) PASS
1833      FORMAT(' TOLERANCE TEST NOT MET IN',I5,' TRIES')
      IF(PASS.GT.10) GO TO 1849
      IF(PPT.EQ.1) GO TO 1803
C  CONVERT TO CONCENTRATION OF PRECIPITATE BEFORE LEAVING
1849    DO 1850 L=N1,N2
        TR=1000.
        DO 1840 K=1,NCM1
        IF(IS(L,K).LE.0) GO TO 1840
        R=BIN(L,K)/IS(L,K)
```

```
      IF(R.LT.TR) TR=R
1840  CONTINUE
      CS(L)=TR
1850  CONTINUE
C  ADJ=-1 CAUSES DISK WRITE OF RESULT FROM SOLVE
      ADJ=-1
      CALL SOLVE
      RETURN
      END
C     LAST LINE OF SPE.FOR
```

```
10 REM "SPEPLOT.BAS"
20 REM: 29 SEP 1988
30 REM: 21 NOV 1986 variable size boxes; labels
40 REM: 8 JAN 87 RECOGNIZES PHINITIAL, (PHINC NOT THOROUGHLY TESTED)
50 REM: AUG 1987:Linear interpolation; plotting of ppt: Use SPE.FOR
60 REM: SEPT 87:PUTS ON LABELS; (PHINC BEING USED) ;(RENUMBERED)
70 PRINT " LAST DATA STATEMENT IS LIST TO SKIP: # FOLLOWED BY LIST (3080)"
80 INPUT "resolution sd? (default=5)";SD
90 IF SD=0 THEN SD=5
100 INPUT "density den? (default=1)";DEN
110 IF DEN=0 THEN DEN=1
120 INPUT "FILENAME=?";FILE$
130 IF FILE$="" THEN FILE$="FOR007.DAT"
139 FOR J=1 TO 9:ORD(J)=J:NEXT J
140 INPUT "L M H label(L M H)    M L H     L H";LMH$
150 IF LMH$="" THEN LMH$="L M H" 'standard species definition heading
160 IF LMH$="L M H" THEN ORD(1)=2:ORD(2)=3:ORD(3)=1
165 IF LMH$="L F H" THEN ORD(1)=3:ORD(2)=1:ORD(3)=2 'special case
170 IF LMH$="M L H" THEN ORD(1)=1:ORD(2)=3:ORD(3)=2 'label order
175 IF LMH$="L H" THEN ORD(1)=2:ORD(2)=1
176 IF LEN(LMH$)>5 THEN ORD(1)=3:ORD(2)=4:ORD(3)=1:ORD(4)=2 'SPECIAL CASE
180 INPUT "Labels on curves? (DEFAULT=Yes)";LBLFLG$
190 LBLFLG$=LEFT$(LBLFLG$,1):IF LBLFLG$="n" THEN LBLFLG$="N"
200 INPUT "Number of Copies? (DEFAULT=1)";NUMCOP%
210 IF NUMCOP%<=0 THEN NUMCOP%=1
220 REM LPRINT DATE$
230 REM LPRINT TIME$
240 REM L=LEN(LMH$):IF L>5 THEN FOR J=1 TO INT((L+1)/2):ORD(J)=J:NEXT J
250 REM IF L>5 THEN TORD=ORD(L):ORD(L)=ORD(L-1):ORD(L-1)=TORD  'reverse
260 T1$=TIME$
270 READ BOXWID,BOXHEI:  ' BOX DIMENSIONS IN INCHES
280 DATA 6,4
290 READ THICK :'BORDER LINE THICKNESS IN INCHES
300 DATA .02
310 XIN=2:XINC=1
320 NTOP=1*300  :'OFFSET FROM TOP
330 NLEF=1*300 : 'OFFSET FROM LEFT
340 WIDTH LPRINT 255
350 READ TOP$,SIDE$,BOT$:'DEFAULT LABELS
360 DATA "TEST SYSTEM","%","-LOG[H+]"
370 GOSUB 3010  'SPECIES SKIPPING
380 DEFSTR P
390 DIM P(256)
400 FOR I=0 TO 255
410 P(I)=CHR$(I)
420 NEXT
430 LPRINT P(27);P(38);P(108);NUMCOP%;P(88);
440 GOSUB 1160: ' READ LABELS AND PARAMETERS
450 GOSUB 530 'MAKE BOX
460 GOSUB 1290 'READ PERCENT DATA & PLOT
470 GOSUB 2310 'READ DISK SPECIES AND COMPONENTS
480 IF LBLFLG$="N" THEN 500
490 GOSUB 1960 'LABELS SECTION
500 GOSUB 2460' FINISH OUT TABLES
510 LPRINT P(12):STOP
520 '
530 REM: BOX
540 NOTIKX=INT(XPHF-XPHI)  'X-axis tick mark count
```

```
550 NHR=INT(BOXWID*300)+INT(THICK*300):  'BOX WIDTH CORRECTED FOR THICKNESS
560 NVR=INT(THICK*300):  'LINE THICKNESS
570 LPRINT P(27);P(42);P(99);NHR;P(97);NVR;P(66); :'COMBINED DEFINE RULE
580 LPRINT P(27);P(42);P(112);NTOP;P(121);NLEF;P(88);:' COMBINED CURSOR
590 LPRINT P(27);P(42);P(99);P(48);P(80);:'PRINT RULE
600 NBOT=NTOP+BOXHEI*300: 'INCHES TO BOTTOM OF BOX
610 NT=NTOP+INT(THICK*300): ' TOP OFFSET CORRECTED FOR LINE THICKNESS
620 LPRINT P(27);P(42);P(112);NBOT;P(89);:'CURSOR TO BOTTOM OF BOX
630 LPRINT P(27);P(42);P(99);P(48);P(80);:'PRINT RULE
640 LPRINT P(27);P(42);P(112);NT;P(121);NLEF;P(88);:' COMBINED CURSOR
650 NHR=INT(THICK*300) 'SET VERTICAL LINE THICKNESS
660 NVR=BOXHEI*300  '  SET VERTICAL LINE LENGTH
670 LPRINT P(27);P(42);P(99);NHR;P(97);NVR;P(66); :'COMBINED DEFINE RULE
680 LPRINT P(27);P(42);P(99);P(48);P(80);:'PRINT RULE
690 NRIGHT=NLEF+BOXWID*300: 'COMPUTE RIGHT SIDE OF BOX
700 LPRINT P(27);P(42);P(112);NRIGHT;P(88);:'MOVE WIDTH OF BOX RIGHT
710 LPRINT P(27);P(42);P(99);P(48);P(80);:'PRINT RULE
720 REM TICK MARKS ON LEFT SIDE
730 LPRINT P(27);P(42);P(112);NTOP;P(121);NLEF;P(88);:' COMBINED CURSOR
740 NHR=INT(.125*300):  'TICK MARK LENGTH
750 NVR=INT(THICK*300):  'TICK MARK THICKNESS
760 LPRINT P(27);P(42);P(99);NHR;P(97);NVR;P(66); :'COMBINED DEFINE RULE
770 FOR I=0 TO 10
780 NT=NTOP+INT(I*BOXHEI*300/10) 'WHERE TO PUT TICK MARK
790 NL=NLEF 'WHERE TO PUT TICK MARK
800 LPRINT P(27);P(42);P(112);NT;P(121);NL;P(88);:' COMBINED CURSOR
810 LPRINT P(27);P(42);P(99);P(48);P(80);:'PRINT RULE
820 A$=RIGHT$(STR$(100-10*I),3)
830 NT=NT+15 'ADDITIONAL OFFSET DOWN FOR CENTERING NUMBERS
840 NL=NLEF-4*30 'WHERE TO PUT NUMBER
850 LPRINT P(27);P(42);P(112);NT;P(121);NL;P(88);:' COMBINED CURSOR
860 IF A$=" 0" THEN A$=" "+A$
870 LPRINT A$;
880 NEXT
890 NHR=INT(THICK*300):  'TICK MARK THICKNESS
900 NVR=INT(.125*300):  'TICK MARK LENGTH
910 LPRINT P(27);P(42);P(99);NHR;P(97);NVR;P(66); :'COMBINED DEFINE RULE
920 FOR I=0 TO NOTIKX
930 NT=NTOP+INT(BOXHEI*300)-NVR 'WHERE TO PUT TICK MARK
940 NL=NLEF+INT(BOXWID*I*300/NOTIKX) 'WHERE TO PUT TICK MARK
950 LPRINT P(27);P(42);P(112);NT;P(121);NL;P(88);:' COMBINED CURSOR
960 LPRINT P(27);P(42);P(99);P(48);P(80);:'PRINT RULE
970 XIN=XPHI 'INITIAL PH  (VALUE FROM FILE)
980 XINC=INT(XPHINC*10+.00001) ' FROM FILE
985 XINC=1  'override above stmnt. xinc should be 1 pH value/interval
990 A$= STR$(INT(XIN+XINC*I))
1000 A$=RIGHT$(A$,3)
1010 NT=NT+90 'ADDITIONAL OFFSET DOWN FOR PH
1020 NL=NL-42 'ADDITIONAL OFFSET FOR CENTERING PH
1030 LPRINT P(27);P(42);P(112);NT;P(121);NL;P(88);:' COMBINED CURSOR
1040 IF LEN(A$)>2 THEN A$=RIGHT$(A$,2):NL=NL+20:GOTO 1030: '2-DIGIT NUMBERS
1050 LPRINT A$;
1060 NEXT
1070 NT=NTOP+INT(BOXHEI*300/2) 'WHERE TO PUT SIDE LABEL
1080 NL=NLEF-150 'WHERE TO PUT SIDE LABEL
1090 LPRINT P(27);P(42);P(112);NT;P(121);NL;P(88);:' COMBINED CURSOR
1100 LPRINT SIDE$;
1110 NT=NTOP+INT(BOXHEI*300)+120 'WHERE TO PUT BOTTOM LABEL
1120 NL=NLEF+INT(BOXWID*300/2.2-LEN(BOT$)/2*30) 'FROM LEFT SIDE
1130 LPRINT P(27);P(42);P(112);NT;P(121);NL;P(88);:' COMBINED CURSOR
1140 LPRINT BOT$;
1150 RETURN 'END OF BOX ROUTINE
1160 REM READ INITIAL DISK DATA
1170 OPEN "I",#1,FILE$
1180 PRINT "READING THE DISK FILE"
1190 INPUT #1, TOP$
1200 NT=NTOP-INT(.25*300) 'WHERE TO PUT TOP LABEL
1210 NL=NLEF+30 'WHERE TO PUT TOP LABEL
1220 LPRINT P(27);P(42);P(112);NT;P(121);NL;P(88);:' COMBINED CURSOR
1230 LPRINT TOP$;
1240 INPUT #1, XPHI,XPHF,XPHINC
1250 PRINT XPHI,XPHF,XPHINC
1260 INPUT #1, N,NS,NC
1270 PRINT N,NS,NC
1280 RETURN 'GO BACK
1290 REM READ PERCENT DATA
1300 NPTS=120
1310 DIM A%(NPTS,N)
```

```
1320 REM     DEF FN Y(X)=C3(D1)*Y3+C2(D1)*Y2+C1(D1)*Y1
1330 DEF FN Y(X)=Y1+(Y2-Y1)*D1/D 'LINEAR INTERPOLATION
1340 DIM C1(100),C2(100),C3(100) 'COEFFICIENTS OF INTERPOLATION EQUATION
1350 D=INT(BOXWID*300/NOTIKX*XPHINC) 'NUMBER POINTS PER INTERVAL
1360 FOR D1=0 TO D
1370 C3(D1)=(D1-D)*D1/D^2:C2(D1)=D1*(3*D-2*D1)/D^2:C1(D1)=((D1-D)/D)^2
1380 NEXT D1
1390 FOR I= 0 TO NPTS
1395 PRINT I; 'point # echo'
1400 FOR J= 1 TO N
1410 INPUT #1,A
1420 IF A>32767 THEN A=32767
1440 IF A<0 THEN 1470
1450 A%(I,J)=A
1460 NEXT:NEXT
1470 NPTS=I-1:IF NPTS>100 THEN NPTS=100 'temporarily expedient
1480 NHR=INT(THICK*300+.5):  'TICK MARK WIDTH
1490 NVR=INT(THICK*300+.5):  'TICK MARK LENGTH
1500 NVR=3:NHR=2 'TRY THIS FIXED QUANTITY
1510 LPRINT P(27);P(42);P(99);NHR;P(97);NVR;P(66); :'COMBINED DEFINE RULE
1520 FOR J=1 TO N
1530 IF J=SKIP(J) THEN 1930
1540 PRINT "SPECIES BEING PRINTED";J,"TIME=";TIME$
1550 D0=0
1560 FOR I=0 TO NPTS-1
1570 PRINT I;
1580 IF I>NPTS-1 THEN 1610 'END OF GRAPH CONSIDERATION
1590 IF A%(I,J)=0 AND A%(I+1,J)>0 THEN A%(I,J)=1' SLOPING UP FROM 0%
1600 IF A%(I,J)>1 AND A%(I+1,J)=0 THEN A%(I+1,J)=1' SLOPING DOWN INTO 0%
1610 Y1=A%(I,J)
1620 IF Y1>999 AND A%(I+1,J)<=999 THEN 1640 'SLOPING DOWN FROM ABOVE 100 %
1630 IF Y1<1 OR Y1>999 GOTO 1920
1640 Y1=Y1*300
1650 Y2=A%(I+1,J)
1660 Y2=Y2*300
1670 DY=ABS(Y2-Y1)
1680 REM SD=7 'SD=RESOLUTION FACTOR replaced by input on top
1690 Y3=A%(I+2,J)*300
1700 DOTS=DY*BOXHEI/1000  '# dots in DY
1710 IF DOTS=0 THEN NVR=SD:NHR=D:GOTO 1790
1720 IF DOTS/D>=1 THEN 1760  'strech rule according to slope
1730 NVR=SD:NHR=INT(SD*D/DOTS) 'set segment lengths
1740 IF NHR>D THEN NHR=D
1750 GOTO 1790
1760 NHR=INT(SD)
1770 NVR=INT(DY*BOXHEI/1000/D*SD+.5)
1780 IF NVR<SD THEN NVR=SD
1790 NVR=NVR+DEN:NHR=NHR+DEN
1800 IF NHR=NHRP AND NVR=NVRP THEN 1830 'save laser memory
1810 LPRINT P(27);P(42);P(99);NHR;P(97);NVR;P(66); :'COMBINED DEFINE RULE
1820 NHRP=NHR:NVRP=NVR  'store previous rules
1830 FOR D1=D0 TO D-1 STEP NHR-DEN  'CORRECTED FOR FATTENING BY DEN
1840 Y=FN Y(X)
1850 NT=NTOP+INT(BOXHEI*300)-INT(BOXHEI*.001*Y+.5)
1860 IF NT<NTOP THEN 1900 'TEST FOR PROTRUSION
1870 NL=NLEF+INT(D*I+.5)+D1
1880 LPRINT P(27);P(42);P(112);NT;P(121);NL;P(88);:' COMBINED CURSOR
1890 LPRINT P(27);P(42);P(99);P(48);P(80);:  'PRINT RULE
1900 NEXT
1910 D0=D1-D
1920 NEXT
1930 NEXT
1940 REM LPRINT CHR$(12) ********************
1950 RETURN 'FINISHED PLOTTING LINES
1960 REM LABELS SECTION
1970 F=100*10   'CORRESPONDS TO 100% (TOP OF GRAPH)
1980 FOR J=1 TO N 'SCAN THRU SPECIES
1990 IMAX=0:IMIN=0
2000 IF J=SKIP(J) THEN 2280
2010 FOR I=1 TO NPTS-1
2020 IF A%(I,J)>A%(I-1,J) AND A%(I,J)<F THEN IMAX=I
2030 NEXT I
2040 IF IMAX>1 THEN 2100
2050 IF A%(2,J)>0 AND A%(2,J)<F THEN IMAX=2 ELSE 2070
2060 GOTO 2100
2070 FOR I=2 TO NPTS-1
2080 IF A%(I,J)<A%(I-1,J) AND A%(I,J)<F*.95 THEN IMAX=I:GOTO 2100
2090 NEXT I
```

```
2100 NL=NLEF+INT(D*(IMAX-1)+.5)
2110 NT=NTOP+INT(BOXHEI*300)-INT(BOXHEI*.001*A%(IMAX,J)*300)
2120 IF A%(IMAX,J)>50*10 THEN NT=NT+50+SD 'ABOVE 50 % PRINT BELOW CURVE
2130 IF A%(IMAX,J)<=50*10 THEN NT=NT-SD 'BELOW 50 % PRINT ABOVE CURVE
2140 IF J>(N-NSP) THEN JJ=J+NS-N ELSE JJ=J 'SKIP UNPRINTED LABELS
2150 J$="":L=0
2160 FOR J2= 1 TO NC  'COMPOSE LABELS
2170 J1=ORD(J2): IF IS%(J,J1)=0 THEN 2240
2180 J$=J$+MID$(LMH$,2*J1-1,1) :L=L+1
2190 IF IS%(J,J1)<0 OR IS%(J,J1)>9 THEN M=2 ELSE M=1
2200 IF IS%(J,J1)=1 THEN 2240      'UNITY SUBSCRIPT
2210 REM  J$=STR$(JJ):L=LEN(J$):J$=MID$(J$,2,L-1) 'CONVER # TO ALPHAMERIC
2220 J$=J$+P(27)+P(42)+P(112)+STR$(NT+20)+P(89)+RIGHT$(STR$(IS%(J,J1)),M)
2230 J$=J$+P(27)+P(42)+P(112)+STR$(NT)+P(89):L=L+M
2240 NEXT J2
2250 IF IMAX>=L AND IMAX<NPTS-L THEN NL=NL-INT(D*L/2)
2260 IF IMAX>NPTS-L THEN NL=NL-INT(D*L)
2270 LPRINT P(27);P(42);P(112);NT;P(121);NL;P(88);J$; :'LABEL "J$"
2280 NEXT J
2290 REM END OF LABEL SECTION
2300 RETURN
2310 DIM T(NC),NAM$(NC)
2320 DIM IS%(NS,NC+1)      '+1 TO PREVENT DIMENSION OVERFLOW
2330 DIM LBETA(NS)
2340 DIM MAX(NS),XPHMAX(NS)
2350 FOR I=1 TO NS         'READ DISK FOR SPECIES
2360 INPUT#1,X,LBETA(I)
2370 FOR L=1 TO NC
2380 INPUT #1,IS%(I,L)
2390 NEXT L
2400 NEXT I
2410 FOR I=1 TO NC          'READ DISK FOR COMPONENTS
2420 INPUT #1,T(I),NAM$(I)
2430 NEXT I
2440 INPUT #1,NSP 'NUMBER OF SOLUBILITY PRODUCTS
2450 RETURN
2460 NT=NTOP+INT((BOXHEI*300))+120+75
2470 LPRINT P(27);P(42);P(112);NT;P(121);NL;P(88);:'COMBINED
2480 FF=50
2490 LPRINT
2500 LPRINT TAB(FF);"Component  Concentration"
2510 LPRINT TAB(FF);"---------  -------------"
2520 FOR I=1 TO NC 'pad names with blanks
2530 L=LEN(NAM$(I))
2540 FOR J=1 TO 10-L '2 extra spaces
2550 NAM$(I)=NAM$(I)+" "
2560 NEXT J
2570 NEXT I
2580 FOR I=1 TO NC 'print names and concentrations
2590 LPRINT TAB(FF);NAM$(I);T(I)
2600 NEXT I
2610 REM  CONSTANTS AND SPECIES TABLE
2620 PRINT "WAIT, COMPUTING MAXIMUM PERCENTAGES"
2630 FOR J=1 TO N
2640 MX=0:XPX=-1
2650 FOR I=0 TO NPTS
2660 IF A%(I,J)>MX THEN MX=A%(I,J):XPX=I
2670 NEXT I
2680 MAX(J)=MX/10
2690 XPHMAX(J)=XPHI+XPX*XPHINC
2700 IF XPX=-1 THEN XPHMAX(J)=0
2710 NEXT J
2720 NT=NTOP+INT((BOXHEI*300))+120+75
2730 NL=0
2740 LPRINT P(27);P(42);P(112);NT;P(121);NL;P(88);:'COMBINED
2750 LPRINT
2760 REM IF NC=2 THEN LMH$="L H"
2770 TEN=5
2780 LPRINT TAB(TEN);" N  Log(Beta)  Max %  at pH  ";LMH$
2790 LPRINT TAB(TEN) "--- ---------  -----  -----  ";
2800 FOR I=1 TO LEN(LMH$)/2+1'extend underline
2810 LPRINT "- ";
2820 NEXT I
2830 LPRINT   'cr
2840 FOR J=1 TO NS
2850 LPRINT TAB(TEN);
2860 LPRINT USING "\  \";STR$(J);
2870 LPRINT USING "\         \";STR$(LBETA(J));
```

```
2880 LPRINT USING "\      \";STR$(MAX(J));STR$(XPHMAX(J));
2890 FOR K=1 TO NC
2900 LPRINT STR$(IS%(J,K));
2910 NEXT K
2920 LPRINT
2930 NEXT J
2940 TS1=VAL(MID$(T1$,7,2)):T1=VAL(MID$(T1$,4,2))*60+TS1
2950 TS2=VAL(MID$(TIME$,7,2)):T2=VAL(MID$(TIME$,4,2))*60+TS2
2960 TIMEX=INT((T2-T1)/60*100)/100:IF TIMEX<0 THEN TIMEX=60+TIMEX
2970 PRINT "TOTAL TIME WAS ";TIMEX;"MINUTES. RESOLUTION FACTOR SD=";SD;
2980 PRINT "  "TIME$;" ";DATE$;
2990 LPRINT P(12);
3000 STOP
3010 REM REGARDING SPECIES SKIP
3020 READ NSKIP
3030 DIM SKIP(100)
3040 FOR J=1 TO NSKIP
3050 READ  SKIP: SKIP(SKIP) =SKIP
3060 NEXT
3070 RETURN
3080 DATA 0
3090 DATA   2,3,4,5,6,7,8,9,10,11,12,13,14,15
```

APPENDIX IV

SCIENTIFIC AND TECHNICAL FIELDS REQUIRING STABILITY DATA

Stability constants are needed to determine the nature of the metal complexes formed under a wide variety of conditions for many applications in diverse areas of science and industry. The following are a few examples of such applications.

ANALYTICAL CHEMISTRY

Quantitative information of the extent of complex formation and the stabilities of complexes is essential for developing methods for the separation of metals from mixtures, and for their determination by electrochemical techniques. Separation methods such as solvent extraction, chromatography, and ion exchange frequently depend on the relative stabilities of metal complexes in aqueous solution in contact with a solid phase or an organic solvent.

BIOLOGICAL SYSTEMS

Accurate values of stability constants are needed to understand biological activity of metals, and metal transport in both plants and animals. Some organisms produce chelating agents to obtain essential metals from their environment (e.g., microbial siderophores). The processes in higher forms of life are much more complex because of the large numbers of carriers (complexing agents) present. Many medical applications are based on complex formation to remove excess toxic metals (e.g., EDTA, Desferal) or to insert and transport metals such as radioactive gallium, indium, and technetium for diagnostic purposes.

COMPLEXATION OF METALS IN AGRICULTURE

Stability constant data are needed to determine the degree of complex formation of metals in soils, a factor that greatly affects their toxicities and availability to plants. The transport of essential trace metals in soils is dependent on complex formation with the natural complexing agents present. Synthetic chelating agents are now employed on a large scale in plant nutrition to form soluble metal complexes which are available to plants. Stability constants are employed to determine the effectiveness of these artificial metal carriers.

CORROSION AND CHEMICAL CLEANING

The corrosion of metals and the cleaning of corrosion products from metal surfaces in both fossil fuel and nuclear steam generators are processes that are profoundly dependent on the metal complexes formed in solution. Effects on metal ion concentrations achieved by the addition of chelating agents is an important factor in the maintenance of steam generators and the stability constants of the metal chelates formed offer the most effective way of assessing metal ion levels in these systems.

DETERGENTS

Large scale use of complexing agents (e.g., NTA, polyphosphates) in detergents requires knowledge of stabilities of metal complexes formed, not only for the detergent action itself, but for the effect of these complexing agents on sewage treatment and on the environment when the detergent solutions are released. Recently new chelating agents having high metal ion affinity are being tested as detergent additives because they produce favorable results in the cleaning process. This action is being attributed to the stabilities of complexes formed with transition metals in soil deposits, which otherwise would result in the formation of stain-producing metal deposits on fabrics.

ELECTROPLATING

The presence of chelating agents in electroplating solutions strongly influences the rate and efficiency of the process and the nature of the deposit. The electrodeposition of metals is controlled by the stabilities of complexes formed in the plating solutions, which frequently contain a number of complexing agents. The use of new types of metal complexes is important for replacement of cyanide baths, which are now considered too toxic for large scale use. The stability constants of metal complexes provide one of the best approaches to the understanding of complex multicomponent electroplating solutions.

ENVIRONMENTAL PROBLEMS

Of the 50,000 organic compounds manufactured in the U.S.A., toxicity data are available on less than one percent, although most of them eventually find their way into the environment. The situation is further complicated by the fact that some of these compounds complex metal ions that are naturally present in environmental systems, or added through contamination, and the toxicities of the complexes formed vary greatly with the nature of the metal ion and ligand. Further complications arise through natural processes, such as hydrolysis, air oxidation (via microbial and enzyme-catalyzed oxidation, and photo-catalyzed oxidation) that convert organic pollutants into complexing agents for metal ions. The toxic effects of some metal ions are increased when the complex is mobile and sufficiently labile to exchange with other receptors in biological systems. On the other hand, formation of a very stable, and hence inert, complex may nearly completely detoxify the metal ion.

The essential first step in estimating toxicities of complex environmental systems is to determine the distribution of metals among the large number of ligands present, a process that requires a knowledge of the stability constants involved. This requires a critical evaluation of the literature to identify the reliable equilibrium constants, and for the estimation (or experimental determination) of those that have not been measured previously.

GEOSCIENCES

The natural processes involving the leaching of minerals by underground waters, metal transport processes, and the deposition and growth of mineral deposits, involve complex formation with the natural metal carriers present. Stability constants of the complexes in solution are important factors in understanding these processes.

MEMBRANE RESEARCH

The development of synthetic membranes is an important area of research being conducted in both industrial and university laboratories. Many of the membranes being designed are based on models that utilize our current knowledge of metal-ligand affinities. Thus a membrane channel may contain certain functional groups that combine with metal ions in such a manner as to promote selective migration through the channel. Conversely, the channels may have stable metal complexes fixed in place which combine loosely with small molecules such as CO, CO_2, O_2, so as to promote their transport.

METALLURGY

Extractive metallurgy frequently involves the action of aqueous solutions of complexing agents on ores. Secondary separations, including solvent extraction and ion exchange also use complexing agents which combine selectively with metal ions. Stability constants are needed for a full understanding of the chemical reactions involved.

OCEANOGRAPHY

The transport of metals in marine systems, as well as the solubilization and deposition of metal compounds in the oceans, depends to a large extent on complex formation. Many trace metals are carried by organic material through complex formation.